THERMOPLASTIC PIPING FOR POTABLE WATER DISTRIBUTION SYSTEMS

BRAB-Federal Construction Council
Technical Report No. 61

Prepared by Task Group T-52
of the
Federal Construction Council
Building Research Advisory Board
Division of Engineering
National Research Council

NATIONAL ACADEMY OF SCIENCES
Washington, D. C.
1971

NOTICE: The study reported herein was undertaken by the Federal Construction Council of the Building Research Advisory Board under the aegis of the National Research Council. The Federal Construction Council program has the express approval of the Governing Board of the National Research Council. Such approval indicated that the Board considered the program of the Federal Construction Council as nationally significant, that elucidation and solution of the problems within the program required scientific or technical competence, and that the resources of the National Research Council were particularly suitable to the conduct of the program. The institutional responsibilities of the National Research Council were discharged in the following manner:

The members of the study committee were selected for their individual scholarly competence and judgment with due consideration for the balance and breadth of disciplines. Responsibility for all aspects of this report rests with the study committee, to whom sincere appreciation is expressed.

Although the reports of study committees are not submitted for approval to the Academy membership or to the Council, each report is reviewed by a second group of scientists according to procedures established and monitored by the Academy's Report Review Committee. Such reviews are intended to determine, *inter alia,* whether the major questions and relevant points of view have been addressed and whether the reported findings, conclusions, and recommendations arose from the available data and information. Distribution of the report is permitted only after satisfactory completion of this review process.

This report was prepared under Contract No. NBS-1096-71 between the National Academy of Sciences and the National Bureau of Standards and is published by the National Academy of Sciences with the concurrence of the Agency Members of the Federal Construction Council. Requests for permission to reprint or quote extensively from this report, from individuals or organizations other than agencies of the United States Government, should be directed to the National Academy of Sciences.

Available from

Printing and Publishing Office
National Academy of Sciences
2101 Constitution Avenue
Washington, D.C. 20418

By supporting contract agreement, Federal agencies wishing copies of this report are entitled to such copies on request to: Building Research Advisory Board, Division of Engineering, National Research Council, Washington, D.C. 20418

Inquiries concerning this publication should be addressed to: The Executive Director, Building Research Advisory Board, Division of Engineering, National Research Council, 2101 Constitution Avenue, NW, Washington, D.C. 20418

International Standard Book Number: 0-309-01934-6

Library of Congress Catalog Card Number 77-180651

Printed in the United States of America

FEDERAL CONSTRUCTION COUNCIL
of the
BUILDING RESEARCH ADVISORY BOARD

The Federal Construction Council serves as a planning, coordinating, and operating body to encourage continuing cooperation among Federal agencies in advancing the science and technology of building as related to Federal construction activities.

In this pursuit, its specific objectives include: Assembly and correlation of available knowledge and experience from each of the agencies; elimination of undesirable duplication in investigative effort on common problems; free discussion among scientific and technical personnel, both within and outside the Government, on selected building problems; objective resolution of technical problems of particular concern to the Federal construction agencies; and appropriate distribution of resulting information.

The Council as such comprises eleven members appointed by the BRAB Chairman from among BRAB membership, plus one member from the senior professional staff of each of the supporting Federal agencies (currently ten), also appointed by the BRAB Chairman on nomination from the individual agencies; all appointments are subject to approval by the President of the National Academy of Sciences.

The Council directs the conduct of technical investigations and surveys of practice, holds symposium/workshops, arranges for interchanges of information and for monitoring of research and technical projects.

COUNCIL MEMBERSHIP—1970-1971

Chairman
William L. McGrath, *Assistant to the Chairman*
Carrier Corporation, Syracuse, New York

Members

Virgil W. Anderson, *Director,* Facilities Engineering Division, National Aeronautics and Space Administration, Washington, D. C.

Stewart D. Barradale, *Manager Construction Research Department,* Research Division, Weyerhaeuser Company, Seattle, Washington

George B. Begg, Jr., *Director,* Program Management Division, Public Buildings Service, General Services Administration, Washington, D. C.

Jack B. Blackburn, *Professor and Head,* Department of Civil Engineering, Kansas State University, Manhattan, Kansas

Herbert Brezil, *Assistant to the Assistant Director,* Division of Construction, United States Atomic Energy Commission, Washington, D. C.

Patrick Conley, *Vice President,* The Boston Consulting Group, Inc., Boston, Massachusetts

Cameron L. Davis, *President,* Miller-Davis Company, Kalamazoo, Michigan

George E. Distelhurst, *Director,* Research Staff, Office of Construction, Veterans Administration, Washington, D. C.

Leander Economides, Economides & Goldberg, Consulting Engineers, New York, New York

Ernest G. Fritsche, *President,* Ernest G. Fritsche and Company, Columbus, Ohio

William B. Holmes, Office of Design, Facilities Department, United States Postal Service, Washington, D. C.

T. W. Mermel, *Assistant to the Commissioner—Research,* Bureau of Reclamation, Department of the Interior, Washington, D. C.

Louis A. Nees, *Associate Deputy Director for Construction,* Directorate of Civil Engineering, United States Air Force, Washington, D. C.

George H. Nelson, P.E., Atlanta, Georgia

Nyal E. Nelson, *Chief, Specifications & Estimates Branch,* Office of the Chief of Engineers, Department of the Army, Washington, D. C.

Joseph A. Rorick, *Director of Design and Engineering,* Real Estate & Construction Division, IBM Corporation, White Plains, New York

Herbert H. Swinburne, FAIA, *Partner,* The Nolen and Swinburne Partnership, Architects, Engineers, Planners, Philadelphia, Pennsylvania

Alfred W. Teichmeier, *President, Diversified Products Division,* U. S. Plywood-Champion Papers, Inc., New York, New York

Harry E. Thompson, *Deputy Chief,* Building Research Division, National Bureau of Standards, Washington, D. C.

Richard H. Welles, *Director, Specifications and Cost Division,* Naval Facilities Engineering Command, Washington, D. C.

FEDERAL CONSTRUCTION COUNCIL

TASK GROUP T-52

THERMOPLASTIC PIPING FOR

POTABLE WATER DISTRIBUTION SYSTEMS

While the Federal Construction Council itself has overall responsibility for its technical programs, specific projects such as this are carried out under the direction of appointed Task Groups of engineers, architects, or scientists, each possessing qualifications in some phase of the subject under consideration. Each Task Group member serves as a specialist in his field or as a generalist in the problem area, not as a spokesman for or a representative of his own agency or any other organization with which he may be associated.

At the request of the Council, the following persons were designated by the various agencies to organize and direct this study:

Irving V. Bloom, Materials Engineer, Naval Facilities Engineering Command, Department of the Navy
Norman B. Coder, General Engineer, Department of the Air Force
William J. Downing, Plumbing Chief, Mechanical and Electrical Engineering Division, General Services Administration
W. R. Henry, Mechanical Engineer, Bureau of Facilities, U.S. Postal Service
Edwin L. Hockman, Bureau of Water Hygiene, Environmental Control Administration, Department of Health, Education, and Welfare
Elmer E. Jones, Agriculture Engineer, Agriculture Research Service, Department of Agriculture
Robert S. Madancy, Sanitary Engineer, Federal Water Quality Administration, Environmental Protection Agency
William A. Schmidt, Mechanical Engineer, Office of Construction, Veterans Administration
Fredrick W. Sedgwick, Mechanical Engineer, Housing Assistance Administration, Department of Housing and Urban Development
Arthur A. Thue, Chemical Engineer, Office of the Chief of Engineers, Department of the Army

BRAB SUPPORTING STAFF

Donald M. Weinroth, Professional Consultant
Henry A. Borger, Program Manager, Federal Construction Council
Roy A. Dunham, Staff Editor
James R. Smith, Assistant Director--Technical Operations

FOREWORD

The development and acceptance of innovations in methods and materials are essential to the continued growth of the building industry. As necessary criteria are developed for ensuring suitability of such methods and materials as replacements for traditional methods and materials, it is of vital importance that such criteria be neither unrealistically stringent nor unwarrantedly permissive; further, since criteria reflect the current state of the art, periodic reevaluations should be anticipated.

The increasing use of thermoplastic piping by the building industry has created an urgent need for criteria to judge its adequacy. Criteria for its proper installation also are needed, because its physical and structural properties are different from those of piping made of traditional materials.

As a result, the Federal Construction Council requested Task Group T-52 to develop criteria for the application of thermoplastic piping to potable water distribution systems. This report presents the criteria developed by the task group, together with conclusions and recommendations and other related information. A previous report prepared by Task Group T-52 is Federal Construction Council Technical Report No. 52, Rigid Thermoplastic Pipe and Fittings for Residential Drain-Waste and Vent Systems.

This report has been reviewed and approved by the Federal Construction Council, and, on the recommendation of the Council, the Building Research Advisory Board has approved this report for publication.

The Board gratefully acknowledges the work of Task Group T-52 and sincerely appreciates the contributions of others to this effort.

JOHN P. GNAEDINGER, Chairman
Building Research Advisory Board

CONTENTS

Section	Page
I. INTRODUCTION	1
A. Purposes of the Report	1
B. Scope of the Report	1
C. Development of the Report	2
D. Organization of the Report	2
II. CONCLUSIONS AND RECOMMENDATIONS	3
A. Conclusions	3
B. Recommendations	3
III. CRITERIA	4
A. Performance Criteria	4
1. General	4
2. Structural	4
3. Mechanical	5
4. Thermal	5
5. Chemical	5
6. Biological	7
B. Installation Criteria	7
1. Color Coding or Marking	7
2. Insulation	7
3. Support	7
4. Procedure	8
5. Joining	9
C. Design Criteria	9
1. Physical Characteristics	9
2. Flexibility	9
3. Hydrostatic Pressure	10
4. Negative Pressure	10
5. Thermal Contraction and Expansion	10

	Page

IV. DISCUSSION. 11

    A. General 11
    B. Suitability of Thermoplastic Piping 12

        1. General Suitability 13
        2. Structural Suitability. 13
        3. Mechanical Suitability. 17
        4. Thermal Suitability 17
        5. Chemical Suitability. 18
        6. Biological Suitability. 19

    C. Design and Installation Considerations. 20

        1. Thermal Expansion 20
        2. Supports and Hangers. 22
        3. Insulation. 24
        4. Underground Burial. 24
        5. Joining 30
        6. Dimensions, Tolerances, and Markings. 32

APPENDIXES

		Page
A.	EVALUATION PROCEDURES.	34
	General.	34
	Test G-1: Pipe and Fittings Inspection.	34
	Structural	35
	Test S-1: Hydrostatic Pressure Test	35
	Test S-2: Crushing Strength Test.	36
	Test S-3: Concentrated Load Test.	39
	Chemical	40
	Test C-1: Waterborne Chemical Test.	40
	Test C-2: Water Contamination Test.	45
	Test C-3: Weathering Resistance Test.	45
	Biological	46
	Test B-1: Fungus Growth Resistance Test	46
B.	STANDARDS FOR PLASTIC PIPING	48
	American National Standards Institute, Inc	48
	American Society for Testing and Materials	48
C.	ADDITIONAL REMARKS ON UNDERGROUND INSTALLATION by Elmer E. Jones, Jr.	51

FIGURES

1. Classification of Underground Burial Conditions. 24
2. Characteristic Failure of Externally Loaded Rigid Pipe . . 25
3. Characteristic Failure of Externally Loaded Flexible Pipe. 26

viii

I
INTRODUCTION

A. PURPOSES OF THE REPORT

The purposes of this report are to present criteria that federal construction agencies can use in determining the suitability of thermoplastic piping for use in potable water distribution systems and to provide guidance for the design and installation of thermoplastic piping.

B. SCOPE OF THE REPORT

The report is concerned only with piping made of thermoplastic materials* intended for use in water mains, water service lines, and building water distribution systems** for all types of structures except those used primarily for high hazard and industrial occupancy.

---

*The two main groups of synthetic organic high polymers that are plastic or formable at some stage are the thermoplastic and the thermosetting types. The thermoplastic materials, unlike the thermosetting materials, consist of long molecular chains (with little or no cross-linking between chains) held together by proximity forces which decrease with temperature increases and are responsible for the material's ability to be melted to liquid form with the application of heat as well as its ability to be reformed into solids with the removal of the heat. The thermoplastics, in common with most plastics, have relatively low densities, relatively low moduli of elasticity, relatively high coefficients of expansion, a tendency to creep and to continue to deform under long-term loading conditions, and low thermal conductivity; they are also not prone to corrode. Typical of the thermoplastic materials are acrylonitrile-butadiene styrene (ABS), polyethylene (PE), polyvinyl chloride (PVC), and cellulose acetate butrate (CAB).

**Definitions of these major components of a potable water distribution system are as follows:

    Water main--the component that conveys water for public use.
    Water service line--the component that conveys water from a water main (or private source of water) to the building being served.
    Building water distribution system--the component that conveys water from the service line to usage points within the building.

Specifically not considered are water transmission lines (i.e., those pipelines conveying water from its source through treatment plants to the point where the water mains begin) and other specialized lines (e.g., separate fire protection systems for buildings) whose designs are ordinarily based upon full and detailed engineering investigation of pipe material properties as well as of piping layout.

The report is not concerned with whether any commercially available thermoplastic pipe, made according to current American Society for Testing and Materials (ASTM) Specifications, Department of Commerce Product Standards, or any other widely recognized criteria, would satisfy the evaluation procedures that are set forth in the report. (Nonetheless, the task group believes that the evaluation procedures presented in Appendix A are realistic and necessary.)

Finally, the report does not contain an evaluation of thermoplastic piping relative to piping made of more traditional materials.

C. DEVELOPMENT OF THE REPORT

In developing this report, the task group relied primarily upon the experience and judgment of its members, supplemented by relevant technical data and other information solicited from the plastics industry, collected through search of the literature, and derived from other knowledgeable persons. The task group did not make field investigations or laboratory tests, but the members did consider the results of investigations and field tests made by others.

D. ORGANIZATION OF THE REPORT

In addition to this Introduction, the report has three main sections: Conclusions and Recommendations (the views of the task group regarding the need for criteria for thermoplastic piping and recommendations as to the use of such criteria); Criteria (statements of criteria grouped as performance criteria, installation criteria, and design criteria); and Discussion (rationale and supporting information upon which the conclusions, recommendations, and criteria are based). Recommended evaluation procedures and a listing of existing standards for plastic piping are contained in appendixes to the report.

II
CONCLUSIONS AND RECOMMENDATIONS

A. CONCLUSIONS

1. Over and above existing industry standards and government specifications, as well as national, regional, and municipal plumbing codes, criteria are needed for determining the acceptability of pipe and fittings manufactured from thermoplastic material for use in potable water distribution systems.

2. Thermoplastic pipe and fittings meeting the criteria set forth in this report can be considered suitable for use in potable water mains, service lines, and building distribution systems.

3. Design and installation of water distribution systems using thermoplastic pipe and fittings require the same full consideration of material properties and loading conditions as are required in the design and installation of systems using pipe and fittings manufactured from traditional materials.

B. RECOMMENDATIONS

1. To be considered acceptable for use in potable water distribution systems, thermoplastic pipe and fittings should meet the performance criteria presented in Section III of this report, as well as the minimum levels of acceptability recommended in the evaluation procedures presented in Appendix A.

2. Thermoplastic pipe and fittings should be installed in accordance with recommendations of the pipe manufacturer, except that the installation criteria presented in Section III of this report should be used in lieu of imprecise or less stringent recommendations of the pipe manufacturer.

3. Design of water distribution systems using thermoplastic pipe and fittings should be based upon the performance and installation criteria and should be accomplished in accordance with the design criteria as set forth in Section III of this report.

III
CRITERIA

The criteria for thermoplastic pipe and fittings presented in this section are grouped as performance criteria, installation criteria, and design criteria. Criteria concerning performance are classified as general, structural, mechanical, thermal, chemical, and biological. Performance criteria assume that installation is made in accordance with recommendations of the pipe manufacturers, or in accordance with criteria presented in this report when those of the manufacturer are ill-defined or less stringent.

A. PERFORMANCE CRITERIA

1. General

Pipe and fittings should have a useful life consistent with the planned economic life of the potable water distribution system of which they are a component part. Specifically, pipe and fittings should have a useful life as follows: (1) 50 years for water mains and water service lines, and for building water distribution systems in other than residential buildings; and (2) 40 years for building water distribution systems in single-family and multifamily residential buildings.

Pipe and fittings should be free of defects and blemishes on inside and outside walls and should be manufactured to acceptable dimensional tolerances. Specifically, they should not contain any blisters, bubbles, cracks, craters, crazings foreign inclusions, pimples, pinholes, pits, or other injurious defects, and should conform to dimensional requirements of recognized national standards.

2. Structural

Pipe and fittings as well as joints[*] should withstand loads imposed during installation and use without damage that impairs performance.[**] Specifically, they should have sufficient strength to withstand an

---

[*]Joints include joints between lengths of pipe, between pipe and fittings, as well as between pipe and fittings made from thermoplastic material and pipe and fittings made from other materials.

[**]Damage that impairs performance of pipe and fittings includes loss of pressure, bursting, cracking, splitting, weepage, seepage, localized ballooning, and excessive creep.

internal hydrostatic pressure equal to or greater than 150 percent of the maximum hydrostatic pressure of the system.*

For pipe 6 in. or larger, the pipe, fittings, and joints intended to be installed below ground should support soil-burial and surcharge loads specified by the piping system designer without sustaining a diametric change in excess of 5 percent. If intended to be installed above ground, they should support external and internal dead and live loads specified by the piping system designer without undergoing a permanent midspan deflection in excess of 0.5 in. If intended to be installed vertically in water wells, they should support external tensile loads of pumps and pumping appurtenances specified by the piping system designer without sustaining elongation in excess of 1 percent.

For pipe less than 6 in., the pipe, fittings, and joints intended to be installed below ground should support a soil burial load of 1,000 psf plus a surcharge load of 300 psf without sustaining a diametric change in excess of 5 percent. If intended to be installed above ground, they should support the dead-load weight of the pipe, the weight of the water being conveyed, plus an external concentrated load of 25 lb. at midspan without undergoing a permanent midspan deflection in excess of 0.5 in. If intended to be installed vertically in water wells, they should support external tensile loads of pumps and pumping appurtenances specified by the piping-system designer without sustaining elongation in excess of 1 percent.

3. Mechanical

Each pipe, fitting, and joint should be capable of operating without leakage loss.

4. Thermal

Pipe, fittings, and joints should withstand rapid temperature changes, and all temperatures encountered in the water being conveyed without resulting occurrence of damage that impairs performance of the pipe, fittings, and joints. Specifically, pipe, fittings, and joints intended for use in cold water service should convey water within a temperature range of $32°$ and $90°$ F. If intended for use in domestic hot water service, they should convey water within a temperature range of $32°$ and $180°$ F. If intended for use in nondomestic hot water service, they should convey water within a temperature range of $32°$ and $205°$ F.

5. Chemical

Pipe, fittings, and joints should be resistant and remain so for their useful life, both to chemicals encountered in the water being conveyed

---

*In general, maximum hydrostatic pressure of the system means the pressure determined by the piping-system designer as being the maximum

as well as to chemicals encountered or normally used in or around installed pipe, fittings, and joints. Specifically, they should be resistant to the concentration of chemicals permitted in potable water by the Public Health Service Drinking Water Standards[*] to the extent that damage that impairs the performance of pipe, fittings, and joints does not occur. If they are intended to be installed below ground they should be resistant to the concentration of calcium, magnesium and sodium salts, iron oxide, manganese, and oxygen, carbon dioxide, hydrogen sulfide, and nitrogen gases that may be encountered in the soil.

Pipe, fittings, and joints should not acquire odors as a result of usage, and should not decrease the potability of water being conveyed below that permitted by the Public Health Service Drinking Water Standards.[*] Specifically, they should not add any of the following substances to the water being conveyed: arsenic, barium, cadmium, chromium (hexavalent), cyanide, fluoride, lead, selenium, or silver.

They should not cause existing concentrations of the following chemicals in the water being conveyed to exceed the limits indicated below (in milligrams per liter):

alkylbenzene sulfonate	0.5
chloride	250.0
copper	1.0
carbon chloroform extract	0.2
iron	0.3
manganese	0.05
nitrate	45.0
phenols	0.001
sulfate	250.0
zinc	5.0

Pipe, fittings, and joints should not cause the water being conveyed to exceed the following limits and thus become offensive to the user's sense of sight, taste, or smell:

turbidity	5 units
color	15 units
threshold odor number	3

In addition, they should not emit toxic fumes as a result of exposure either to water temperatures within the piping or to ambient and other heat sources apt to be encountered around installed piping.

---

pressure to be experienced by the system as a result of pumping pressure and pressure associated with anticipated water surges.

[*]U. S. Department of Health, Education, and Welfare, Public Health Service, Public Health Service Drinking Water Standards, (Washington: U. S. Government Printing Office, latest edition).

6. Biological

Pipes, fittings, and joints should be resistant to attack by microorganisms and should have internal and external surfaces such that no nutrients are provided that would sustain the growth of microorganisms.

B. INSTALLATION CRITERIA

Thermoplastic pipe and fittings satisfying the foregoing criteria for acceptability should be installed in accordance with recommendations of the pipe manufacturer, except that installation, marking, and sizing criteria contained in standards of the American National Standards Institute and the American Society for Testing and Materials should prevail; e.g., ASTM 2774-69, Recommended Practice for Underground Installation of Thermoplastic Pressure Piping.* In lieu of ill-defined or less stringent recommendations, the practices described here should be followed.

1. Color Coding or Marking

If thermoplastic pipe and fittings are to be installed in more than one piping system within the same building water distribution system, then pipe and fittings either should be color coded or otherwise marked so that immediate identification of the proper use of each length of pipe and each fitting is possible. Otherwise, pipe and fittings for all systems should meet the pipe and fittings criteria for the most severe system. All pipe and fittings should also be marked in accordance with the marking system included in the standards of the American Society for Testing and Materials for pipe and fittings.

2. Insulation

Thermoplastic piping should be insulated when installed in areas where the ambient temperature can be below 32° F. or above 90° F. in the case of piping suitable only for cold water usage, or 180° F. in the case of piping acceptable for hot water usage. When insulation is employed, it should conform to the recommendations of the pipe manufacturer.

3. Support

Unless unusual conditions prevail with respect to the supporting of piping, spacing of hangers for horizontal runs of piping should be based on an allowable permanent midspan sag between hangers of 0.5 in. when the piping is conveying water at the maximum anticipated temperature and is subjected to anticipated external loading conditions

---

*The current standards from these two organizations for plastic piping are listed in Appendix B.

(such as high ambient temperatures and loads to be suspended from the piping); however, in no instance should the interval between hangers exceed 8 ft. and hangers should be provided near the end of all branch lines, at all changes in piping direction, and near the end of the run of piping.

Spacing of supports for vertical runs of piping should be such that one support is provided at the base of the run and at least one support is provided per story.

Individual hangers and supports should have a minimum bearing width of 0.75 in.

Piping should not be anchored so securely by a hanger or support that freedom of movement to accommodate thermal expansion or contraction is impaired; when mechanical expansion joints are incorporated in the system layout, these joints should be anchored in such a manner as to prevent motion of the joint itself, but any intermediate support or hanger between such joints should not restrain the movement of the piping.

4. Procedure

Unless unusual conditions prevail, the following installation procedures should be adhered to for all thermoplastic piping having a diameter of 5 in. or more. (For smaller diameter piping--i.e., less than 5 in.--these requirements can be waived.)* **

Ditches (or subditches) for trench or negative projecting embankment installations should be excavated to the minimum width required for proper installation and to the required depth with straight sides below the top of the pipe; the bottom should be smooth, have a uniform grade, and should have a minimum bedding layer of 4 in. compacted, selected, fine-soil materials placed in the bottom. The pipe should be so placed in the ditch that the pipe bears on the bedding throughout the entire length of the pipe barrel (i.e., offsets should be shaped in the bedding to fit joint hubs). The ditch should then be backfilled to a depth of at least 12 in. above the top of the pipe with soil material similar to that of the bedding; care should be taken to compact the backfill thoroughly under the haunches of the piping; the 12 in. of backfill immediately above the pipe should not be compacted.

---

*When soil conditions peculiar to a specific installation require extra precaution, as in the case of expansive clays, the same precautions normally taken with piping of traditional material of manufacture should be followed.

**See Appendix C for additional remarks by Elmer E. Jones, Jr.

For positive projecting embankment installations, the bedding should be so prepared that the piping lies true to grade, and the backfill should be compacted for a width on each side of the pipe equal to two times the outside diameter and to a height of at least 12 in. above the top of the piping; the 12 in. of backfill immediately above the piping should not be compacted.

For all underground installations, backfill and bedding should be of selected fine compactible soil materials free of stones, boulders, sharp objects, or other waste; both backfill and bedding should be compacted at near optimum moisture content in layers not exceeding 6 in. in compacted thickness to a density of at least 90 percent of the maximum density at a compactive effort of 12,400 ft.-lb. per cu. ft.

The pipe crown should be installed below the frost-penetration level anticipated by the designer.

5. Joining

With respect to joining pipe and fittings, threads should not be cut in any pipe having a wall thickness less than Schedule 80.

When insert fittings are used in underground piping, stainless steel clamps should be used. This joining method should not be used when the soil is apt to corrode stainless steel.

When solvent cement is used for joining, a solvent cement meeting nationally recognized standards should be used if such standards exist. Otherwise, the cement recommended by the pipe manufacturer should be used. The solvent cement should be in containers permanently marked with (1) the specification, if any, with which the solvent complies or the type of solvent cement material, (2) plastic materials with which solvent cement can be used, (3) name or trademark of manufacturer, (4) instructions for use, and (5) safety precautions, if any, to be taken while using the solvent cement. Field cementing should be done only on interference-fit joints.

C. DESIGN CRITERIA

1. Physical Characteristics

Because the various types of thermoplastic material have different physical characteristics that can markedly affect the size of pipe used and the design of the piping systems, and hence the cost of the system, piping-system designers should take such differences into account in their design analysis in order to achieve maximum economy and performance.

2. Flexibility

All thermoplastic piping that is currently available commercially can tolerate more than 3 percent diametric change without material damage.

Therefore, it should be considered flexible, and the determination of the amount of soil burial and surcharge loads to be experienced by piping installed underground should be made on the basis of prevailing soil mechanics theory, such as that set forth in Soil Engineering, by Merlin G. Spangler.*

3. Hydrostatic Pressure

Prevailing hydraulic theory is just as applicable to water flow in plastic pipe as it is to water flow in more traditional kinds of pipe. Thus, the determination of the maximum hydrostatic pressure of the system can be made by following prevailing hydraulic theory.

4. Negative Pressure

Because thermoplastic piping in the larger sizes (larger than 4 in.) could collapse under negative system pressure, particular effort should be made to design a system that is never subjected to negative pressure or that can specifically accommodate negative pressures of up to one atmosphere.

5. Thermal Contraction and Expansion

Thermoplastic materials have significantly higher coefficients of thermal expansion than piping material traditionally used. For this reason, design procedures for thermoplastic piping systems should incorporate specific analyses of the consequences of, and the means needed to compensate for, thermal contraction and expansion; in these analyses, unless unusual conditions prevail, it is realistic to assume that the temperature differentials to be accommodated are as follows: (a) for below-ground cold water lines, 100° F.; (b) for above-ground cold water lines, 130° F.; and (c) for all hot water lines, 180° F.

---

*Merlin G. Spangler, Soil Engineering (Scranton, Pa.: International Textbook Co., 1966).

IV
DISCUSSION

A. GENERAL

Until recently, piping for potable water distribution systems has been made from traditional materials, principally materials processed from available natural resources. The acceptability of such piping has been adjudged on the basis of tests conducted in conformance with prescriptions of local codes, various national standards, and government specifications. Such prescriptions have tended to be written on the assumption that only existing traditional materials of manufacture would be used; thus the test and evaluation techniques prescribed have been closely related to the properties of certain traditional materials.

If piping for water mains, water service lines, and building water distribution systems is to be manufactured from different materials, such as man-made thermoplastics, instead of reinforced and unreinforced concrete, asbestos cement, cast iron, steel, copper, and brass, a problem is created. How is the performance of the innovation--until manufactured according to proven specifications and recognized in existing codes--to be evaluated without causing unreasonable predilection toward the innovation, unreasonable injustice to the established products, unreasonable health and safety risk to the general public, and unreasonable economic-technological hazard for the intended purchaser or user of the innovation?

Thermoplastic piping cannot and should not be evaluated strictly according to the letter of prescriptions previously applied to piping made of traditional materials. It is just as unreasonable to insist that thermoplastic materials possess the same physical properties as any one of the more traditional materials as it is to insist that concrete possess the same physical properties as steel. Further, it is just as unreasonable to insist that thermoplastic piping function in precisely the same way as piping of any one of the more traditional materials of manufacture as it is to insist that concrete piping function precisely the same as cast iron piping. However, it is reasonable, and indeed necessary, to insist that thermoplastic piping--when properly manufactured and installed in potable water distribution systems designed to utilize such piping--perform its intended functions without creating unreasonable risks to the public and to users and purchasers.

Translation of this reasonable requirement into rational methods for designing a suitable system to take advantage of thermoplastic piping as well as methods for selecting, specifying, and procuring thermoplastic pipe and fittings on the basis of predicted ability to perform a given function at a particular level of performance is the principal subject of this discussion. Such consideration is given, first, to the function of piping in general and to that of thermoplastic piping in particular, and, second, to the design and installation of piping in general and to that of thermoplastic piping in particular. Concise definition of the function of piping as the conveying of potable water in a manner consistent with the health, safety, and welfare concerns of water users serves little practical purpose, for it does not express the multifaceted conditions that must be satisfied if potable water is to be properly conveyed. To express completely the function of piping in potable water systems requires a series of statements. Such a series, deemed essential to ensuring the proper performance of thermoplastic piping in potable water systems, is set forth in the first part of Section III of this report, subdivided, for convenience, into general, structural, mechanical, thermal, chemical, and biological characteristics.

Once the performance of the piping has been defined by such statements of performance criteria, it becomes necessary to determine how piping should be evaluated for the characteristics in the statements. Ideally, the evaluation should determine the effects of use, wear, and deterioration of the product that occur in service and should provide a nonprejudicial basis for comparing various types of thermoplastic piping and piping of other materials of manufacture. At this juncture the series of evaluation procedures presented in Appendix A departs from the ideal; the evaluation procedures set forth, although they may be applicable to piping of any material of manufacture, have only been considered with respect to evaluating thermoplastic piping.

After stating the criteria in terms of performance characteristics and selecting suitable evaluative techniques, it is necessary to select acceptable levels of performance. Truly valid acceptable levels of performance require giving appropriate weight to user and community needs in addition to determining what combination of technical performance characteristics will satisfy those needs. Although laboratory tests and field evaluations were not a part of the study effort, the levels of performance for thermoplastic piping stipulated for each evaluation procedure set forth in Appendix A merit confidence and provide a rational initial basis for determining the acceptability of thermoplastic piping in potable water distribution systems.

B. SUITABILITY OF THERMOPLASTIC PIPING

To facilitate appreciation of the significance of the performance criteria set forth in Section III and the evaluative techniques and levels of performance presented in Appendix A, the rationale for the criteria and the supporting evaluative techniques and levels of performance are provided.

1. <u>General Suitability</u>

The statements of general performance criteria relate chiefly to the required useful life of the piping. The principal elements relevant to such a requirement are as follows: first, a statement of the required useful life; second, a prohibition against defects and obvious blemishes that would tend to reduce the useful life if they were present; and third, a limit on the variability of dimensions so that proper mating of components in a system is possible.

Although there have been discussions within and outside the building industry in recent years to the effect that current code requirements and design criteria result in excessively long-lived facilities, there is no real promise that shorter lived facilities will become acceptable in the near future. Therefore, the stipulations of intended useful lives for the various classes of water service set forth in the statements of general performance characteristics for thermoplastic pipe and fittings are in accordance with currently accepted schedules of useful life for piping systems. And, although no evaluation procedure specifically examining the useful life capability of pipe and fittings is recommended herein, the issue of useful life is suitably accounted for in Test S-1: Hydrostatic Pressure Test.

Test G-1: Pipe and Fitting Inspection set forth in Appendix A is intended to evaluate the type and number of defects found in new pipe and fittings. The defects cited in the criteria are defined in the evaluation procedure specified in Appendix A.

2. <u>Structural Suitability</u>

The statements of structural performance characteristics include requirements for the piping that relate to the periods of time both before and after installation of the piping. The primary considerations include the capacity of the piping to perform properly following subjection to imposed loads and without endangering or causing damage to adjoining appurtenances. That is, an individual length of pipe and an individual fitting must be capable of being installed and used in such a way as to permit the individual component not only to function under likely conditions of load but also to function without detrimental effect on adjoining lengths of pipe and fittings or on piping-system appurtenances such as kitchen or bathroom fixtures.

In general, piping is exposed to both internal and external loading conditions. Internal loading is principally created by the internal water pressure (the magnitude of which is the sum of the maximum system working pressure--i.e., maximum hydrostatic pressure plus any likely increase resulting from pressure surges associated with starting, stopping, or otherwise modifying water flow). External loading is the resultant of all the various loads that can be applied to the piping from the outside; e.g., loads due to burial of piping (including both soil covering and any static or dynamic loads), loads due to relative motion between components and appurtenances of the piping system, loads due to suspension of objects from exposed piping, and loads due to accidental or abusive treatment of the piping. Concerning internal loading, certain pertinent observations can be made, namely:

    a. Because of the tendency of thermoplastic materials to undergo major physical change and to creep with the application of heat, it is necessary to take explicit notice of the range of water temperatures apt to be encountered.

    b. Because of the relatively low modulus of elasticity of thermoplastic materials at ambient temperatures, it is necessary to take explicit notice of the range of pressure surges likely to be encountered and the elongation of the piping under load.

Test S-1: Hydrostatic Pressure Test included in Appendix A is intended to evaluate the acceptability of thermoplastic piping, taking due account of these necessities. Unfortunately, while it is relatively easy to specify the temperature conditions to be met,* it is

---

*It can be postulated that the range of water temperatures to be encountered in cold-water lines installed above or below ground varies between 32° and 90° F. and that maximum duration of continuous exposure to either temperature extreme being 6 months in any 1 year and the total time of exposure, within an assumed 50-year life of piping, to either temperature extreme being no more than 25 percent of the 50-year period. Current industry practice is to pressure rate cold-water piping based on testing at 73° F., and it is believed that this temperature represents a suitable approximation of the average water temperature to be encountered in cold-water piping. For domestic hot-water lines installed above or below ground, water temperatures can similarly be postulated as being between 120° to 180° F. (with essentially the same duration of exposure as noted) except that it is possible, and indeed probable, that within a 50-year period, some hot water that ranges between 200° and 210° F. momentarily will be encountered in a building water distribution system. For nondomestic hot-water lines, water temperatures on the order of 205° F. are common.

While it is confidently expected that the levels of performance suggested in Appendix A will ensure the acceptability of thermoplastic piping for use in these "normal" situations, the suggested levels are inappropriate for other conditions such as for cold-water piping installed in an uninsulated attic space where the ambient temperature may be in excess of 90° F. for extended periods of time or for hot-water piping carrying water between domestic water

not equitable to specify a common hydrostatic pressure to be accommodated because of the wide variation in system working pressures and pressure-surge increases found in existing piping systems and incorporated in designs for new piping systems.* Rather, it is the responsibility of the system designer to determine the maximum system working pressure--as well as the magnitude of any pressure surges apt to occur--in a particular situation and to judge the acceptability of thermoplastic piping for that situation in light of the design determination. This determination is a significant factor in assessing the acceptability of thermoplastic piping, and it should be made by following prevailing hydraulic theory. Therefore, an extended discussion of how it should be made is not warranted.

Although Test S-1 is based upon ASTM D2837-69, <u>Standard Method for Obtaining Hydrostatic Design Basis for Thermoplastic Materials</u>, one major modification of that standard method is deemed necessary. The test specimen should include joints and fittings as well as lengths of pipe, rather than merely individual lengths of pipe.

The modification of the standard method with respect to the test specimen may well introduce a high degree of scatter in test data because it may be difficult to "standardize" the test specimen (lengths of pipe plus fitting joined in accordance with the recommendations of the pipe manufacturer).** Nonetheless, this modification is desirable and is, in fact, in keeping with the current efforts of the Joint Plastic Pipe Institute - American Society for Testing and Materials Subcommittee on Plastic Pipe and Fittings to develop test methods for evaluating the hydrostatic characteristics of thermoplastic piping systems.

---

   storage tanks and boiler coils. This is not to say that thermoplastic piping is unsuited for such "abnormal" situations but merely to indicate that such particular applications need special investigation and cannot be allowed to determine the general level of acceptability of thermoplastic piping.

*However, it is possible to postulate the range of system working pressures likely to be encountered in each class of service (water mains, water service lines, or building water distribution systems). Typically, water mains and water service lines are operated at pressures between 40 and 250 psi, with the "normal" range being between 60 and 150 psi. Building water distribution lines are operated at pressures between 30 and 150 psi, with pressures most frequently being between 70 to 80 psi.

**A fairly high degree of scatter in test data seems to be inherent in the basic test method (see, for example, Frank W. Reinhart, "Long-Term Hydrostatic Strength Characteristics of Thermoplastics Pipe," <u>Polymer Engineering and Science</u>, Vol. 6, No. 4, Oct. 1966).

With respect to the external loading condition for thermoplastic piping, the most severe and most frequently encountered condition is the result of underground installation of the piping. Although such a statement can be made with respect to piping irrespective of material of manufacture, the statement assumes particular significance for thermoplastic piping because the piping accommodates burial loads in a fashion completely different from that in which such loads are accommodated by piping of traditional materials. Because of this difference in load transfer mechanisms, an extended discussion of the subject is included in this report as part of the discussion about design and installation considerations. However, three peculiarities are noted here:

    a. Piping extruded from thermoplastic materials currently commercially available is flexible piping--i.e., the cross-sectional shape can be so distorted that diametric dimension changes of more than 3 percent do not cause damage to the piping.

    b. Buried flexible piping supports superimposed burial and overburden loads by causing passive lateral soil-resistance pressures to develop in the soil at the sides of the embedment and by using this resistance to develop the elastic ring strength of the piping.

    c. Failure of buried flexible piping generally occurs when the diametric dimensions change by more than 20 percent and the circumferential stress in the wall of the piping due to all loads exceeds the elastic limit of the material.

Test S-2: Crushing Strength Test set forth in Appendix A is intended to evaluate the acceptability of thermoplastic piping pragmatically, taking due note of these three peculiarities. The test procedures impose more stringent loading conditions on the test specimens than would be encountered under actual burial conditions because the specimens are subjected to almost point loads across the vertical diameter without benefit of restraining horizontal forces (such as would develop from passive lateral soil resistance) which would limit horizontal diameter deflection and bring the full elastic ring strength of the piping into play. However, the level of performance suggested in association with this procedure--a level set in order to limit the amount of loss in hydraulic efficiency rather than to provide a large margin of safety against deflections on the order of 20 percent of diametric dimension--all but precludes pipe failure as a result of poor compaction during the extended period of use of the piping.

Test S-3: Concentrated Load Test included in Appendix A is intended to evaluate the acceptability of thermoplastic piping in connection with abusive external loads. Test S-3 is particularly concerned with ensuring that, through the intended useful life of the piping, piping properly installed will not undergo a permanent midspan deflection in excess of 0.5 in. for aesthetic reasons. The recommended requirement for this test of 0.3 in. allowable deflection was selected in part on the assumption that the long-term deflection developed in a test lasting 1 percent of the life of the piping is approximately 60 percent of the deflection that can occur over the entire life of the piping.* It should be noted that, although Test S-3 is based on simulation of realistic conditions, thus far there has been no laboratory verification of the validity of the test procedure or of the ability of any available thermoplastic piping to meet the recommended minimum requirement.

3. Mechanical Suitability

The statement of mechanical performance characteristics is related specifically to the concept that the pipe and pipe fittings must function as part of an overall potable water system--i.e., the pipe and fittings have to receive water from some source, convey the water to some location with minimal loss, and discharge water for some end use. This concept requires that there be reasonable assurance that pipe and fittings can be joined properly to themselves as well as to other piping-system components and appurtenances. However, in view of the installation criteria presented in Section III with respect to joining, it is not necessary to propose a separate evaluation procedure for joint leakage. As an additional safeguard, Test S-1 set forth in Appendix A does include examination of joints at intended system working pressure and temperature and thus would reveal whether particular piping can be satisfactorily joined.

4. Thermal Suitability

The statement of thermal performance characteristics encompass specific consideration of the premise that the pipe and fittings must not be susceptible to damage from either water temperature or changes in water temperature that can occur in the piping system.

No separate test is needed to ensure that pipe and fittings can withstand the temperature of water being conveyed without sustaining damage because Test S-1 set forth in Appendix A includes testing of specimens with water at temperatures representative of service conditions. It should be noted that, although currently available thermoplastic piping meeting ASTM and Product Standard criteria can meet the recommended requirement contained herein for use in cold-water piping systems, no currently available thermoplastic piping has been evaluated to determine its ability to meet the hot-water service requirements contained herein.

---

*This assumption is suggested frequently in regard to creep tests (or long-time tensile strength tests) for determining design stresses for metallic pipe, valves, and fittings for service at elevated temperatures. See, for example, Sabin Crocker, Piping Handbook (New York: McGraw-Hill Book Company, Inc., 1945), pp. 336-346.

5. <u>Chemical Suitability</u>

The statements of chemical performance characteristics relate specifically to four separate considerations.

    a. The inside surface of the pipe and fittings must be resistant to chemicals in the water being conveyed.

    b. The pipe and fittings should not decrease the potability of water being conveyed below that permitted by the <u>Public Health Drinking Water Standards</u>.*

    c. The pipe and fittings should not acquire and retain odors as a result of usage.

    d. The pipe and fittings should not emit toxic fumes as a result of exposure to water temperatures within the piping, ambient temperatures, and other heat sources apt to be encountered around installed piping.

Test C-1: Waterborne Chemical Test set forth in Appendix A is intended to evaluate whether pipe and fittings can withstand chemicals likely to be encountered in the water being conveyed without damage. The test uses as test reagents the concentration of chemicals permitted in potable water by the <u>Public Health Drinking Water Standards</u>,* and incorporates a short-time rupture strength test of the specimens after exposure to the test reagents for 10 days. The recommended requirements, stated in terms of hoop strength of the specimen should ensure continuous satisfactory performance of the piping through its intended life.

Test C-2: Water Contamination Test set forth in Appendix A is intended to evaluate the effect of the piping on the potability of water. The test apparatus and procedures are identical to those reported in National Sanitation Foundation, <u>A Study of Plastic Pipe for Potable Water Supplies</u> (Ann Arbor: National Sanitation Foundation, 1955). The inclusion of the Foundation test herein should not be construed as meaning that thermoplastic piping must bear the Foundation's certification in order to be used in potable water systems.

Test C-3: Weathering Resistance Test set forth in Appendix A is intended to evaluate the ability of above-ground piping to withstand ambient weather conditions without damage. The test and the recommended minimum requirement are predicted on the assumption that the useful life of the piping will be reasonably ensured if ambient conditions do not cause the piping to develop defects and detrimental blemishes set forth in Test G-1 or to lose flexural strength.

---

*U. S. Department of Health, Education, and Welfare, Public Health Service, <u>op. cit</u>.

At this time, separate procedures for evaluating the effect of soils on piping, the tendency of piping to acquire and retain odors, and the tendency of piping to emit toxic fumes as a result of exposure either to water temperatures within the piping or to ambient and other heat sources (e.g., radiant heaters) do not appear warranted.

6. Biological Suitability

The statements of biological performance characteristics are related specifically to two separate considerations:

   a. The pipe and fittings must be resistant to biological attack.

   b. The pipe and fittings should not provide nutrients that would sustain the growth of microorganisms.

A previous report of the Building Research Advisory Board concerned with the use of plastic pipe in drain-waste and vent systems reported that there was insufficient evidence regarding the possibility of damage to thermoplastic pipe and fittings due to attack by vermin to warrant establishing criteria in this area.* Re-examination of this subject as part of this study has similarly failed to reveal any reason for concern.

However, the previous report evinced concern about degradation of thermoplastic materials by fungi; thus a specific test procedure was recommended for ascertaining the suitability of plastic piping for use in drain-waste and vent systems.** Re-examination of this subject as part of this study indicates that actual degradation of plastic pipe due to fungi attack is rare but that there is legitimate concern about plastic pipe becoming unsightly as a result of fungi growth. On this basis, a criterion relating to fungi growth is appropriate. The same procedure to determine that piping made of thermoplastic materials contains no materials or ingredients that would supply nutrients for the growth of fungi, which was recommended in the earlier report, has been repeated herein.

In regard to the specific test method recommended, it should be noted that conclusions regarding the likelihood of a plastic pipe becoming covered with (or degraded by) mycological organisms in the field based on extrapolations of laboratory test results are at best difficult to establish. Therefore, extrapolations of laboratory data must be based on results obtained with proven methods and procedures. For the

---

*Building Research Advisory Board, Rigid Thermoplastic Pipe and Fittings for Residential Drain-Waste and Vent Systems, Federal Construction Council Technical Report No. 52 (Washington: National Academy of Sciences, 1966), pp. 19-20.

**op. cit., pp. 7, 21, 30-32.

purpose at hand, the test method recommended in Technical Report No. 52, which appears in Military Specification MIL-P-82056 (YD), has been effective, giving consistently reliable and reproducible results.

With regard to determining material suitability on the basis of visible evidence of fungus growth on specimens at the termination of the test, it is believed that if the piping material contains no ingredients that provide nutrient for fungus growth, no growth should be evident at the termination of the test.

From time to time during the course of this study, questions were raised regarding the likelihood of buried plastic pipe being attacked by termites. Pursuit of this question with representatives of users of substantial quantities of plastic pipe has failed to introduce any substantive evidence for concern at this time. Moreover, what scientific evidence is available* suggests that termites attack plastic materials such as tape, cable, and cable sheathing having high plasticizer contents, but that termites generally do not attack these same materials or rigid pipe in which either there is no plasticizer or the plasticizer content is low.

C. DESIGN AND INSTALLATION CONSIDERATIONS

Ensuring that thermoplastic piping is acceptable for use in potable water distribution systems does not necessarily ensure that the piping system can function properly; there are several practical design and installation considerations--thermal expansion, hangers and supports, burial, joining, tolerances and dimensions, and material identification--that merit attention. None of these considerations presents problems that cannot be eliminated by following proper procedures, but because each thermoplastic material possesses distinctive characteristics, each must be dealt with on an individual basis.

1. Thermal Expansion

The coefficients of thermal expansion of thermoplastic materials are, comparatively, much greater than those of metals, being on the average at least 5 times (and in some cases as much as 20 times) that of steel. The magnitude of linear expansion associated with such coefficients can be comprehended by noting that a 100-ft. length of Type 1 polyvinylchloride, if installed at 75° F., would expand approximately 2 in. if its operating temperature was 135° F.

---

*See for example, F. J. Gay and A. H. Wetherly, Laboratory Studies of Termite Resistance, IV-The Termite Resistance of Plastics, Division of Entomology Technical Paper No. 5 (Australia: Commonwealth Scientific and Industrial Research Organization, 1962) and R. H. Beal, Testing of Wire Insulating Materials for Resistance to Termites, Second Progress Report, Supplement No. 2, U. S. Department of Agriculture, Forest Service Project FS-SO-2206-6.103 (Washington: U.S. Government Printing Office, 1968).

In general, linear movement resulting from thermal strain does not present a serious problem for either buried piping or piping within a structure. Normally, thermal expansion is not a consideration when the overall temperature range is less than 30° F., even for thermoplastic materials having the highest coefficients of expansion. However, this parameter cannot be discounted in the case of potable water distribution systems because the temperature ranges involved can cause significant thermal expansion (or contraction) that, if completely ignored, can cause failure of the system by developing stresses sufficient to cause rupture of the pipe or to pull a joint loose.

Thermal expansion of piping within a structure can be offset by a variety of design features, each of which is dependent on the geometry and temperature fluctuations of the piping system. The simplest method takes advantage of the inherent flexibility of plastic piping by allowing unrestrained movement at points of directional change. In this manner, the axial stresses resulting from thermal expansion are translated into flexural (bending) stresses of a low order of magnitude as the movement is absorbed by the change in direction of the piping. This technique of offsetting thermal strain is usually adequate for one- and two-family residences of two stories or less. In larger buildings, in which relatively long horizontal and vertical runs of piping may be involved, restraint of the piping at the center of a run can help to distribute the movement uniformly to both ends of a run, where the smaller movements can then be absorbed by directional changes.

When expansion cannot be accommodated by directional changes or when the piping is dimensionally constrained with fixed terminal points because of other design conditions, special provisions to accommodate movement may be required. Under such circumstances, several techniques can be employed--the most common of which is to provide offsets, lyre loops, or U-bends. For large-diameter pipe where lack of space might preclude the use of such bends, mechanical expansion joints available commercially can be used. These mechanical joints consist of a rubber ring inserted in a groove in the socket of the fitting, which allows slippage of the pipe within to absorb linear movement.

Despite the foregoing, piping should not be held securely by any support or hanger, but rather should be secured with strap hangers that allow freedom of movement. When mechanical expansion joints are used, the joint itself should be installed in such a manner as to prevent its movement; however, any intermediate support or fastener between mechanical expansion joints should be of loose fit to allow movement of the piping.

When piping is buried, thermal expansion can normally be accommodated by (1) snaking the piping as it is laid in the trench, (2) backfilling the pipe in the cool of the day, and (3) using offsets and mechanical expansion joints.

Considering the most stringent combinations of minimum and maximum ambient temperatures likely to be involved (0° and 130° F., respectively) as well as the range of water service temperatures likely to be involved, then (1) below-grade piping should be installed to allow for that expansion and contraction that would occur over a temperature differential of 100° F., (2) cold-water lines above grade should be installed to allow for that expansion and contraction that would occur over a temperature differential of 130° F., and (3) hot-water lines above grade should be installed to allow for that expansion and contraction that would occur over a temperature differential of 180° F.

A note of caution should be expressed concerning piping that must be placed along a trench and left exposed for hours during the heat of the day. Many thermoplastic materials expand considerably when exposed to direct sunlight. If the piping is installed and covered while in this expanded state, there can be substantial contraction with serious consequences associated when the pipe reaches the minimum ground temperature anticipated. One way to guard against this effect is to allow the pipe to cool overnight and backfill the trench in the morning. Another way is to put a complete section of piping in service before backfilling. This allows the conveyed water to cool the pipe to a temperature that is within the range to be encountered in service. Such procedures might be required only on an extremely hot day, and the manufacturer should be consulted.

Recommendations of the pipe manufacturer should be sought regarding the total elongation or contraction that would occur over anticipated temperature ranges, because the thermal coefficient of expansion varies significantly for each thermoplastic material. In addition, recommendations of the manufacturer should be solicited and followed concerning the use, location, and anchoring of offsets, U-bends, and rubber ring-type expansion joints, the location of hangers and supports with respect to directional changes of piping within a structure, and methods of checking after installation to ensure that sufficient allowance for thermal strain has been built into the system.

2. Supports and Hangers

Consideration of mechanical and thermal strength requirements for thermoplastic pipe cannot be separated from such design features as the spacing of supports or hangers, because the external load-carrying capability and the amount of sag or deflection that occurs between supports of horizontal runs of piping in buildings at service temperatures are a function of the support spacing.

Even though the weight per lineal foot of thermoplastic piping--including the weight of conveyed water--imposes only nominal loads on individual hangers, the relatively soft texture and low ring-strength of these materials require that hangers with broad, smooth bearing surfaces be used, rather than ones with narrow or sharp-edged contacts, in order to minimize stress concentrations, distortion, and physical damage to the pipe. Although clamp, saddle, friction, or other standard types of hangers can be used, the strap type is preferable. However, no matter which type is used, individual hangers should have a minimum bearing width for the pipe of 3/4 in.

Unless dictated by specific design conditions, hangers should never be tightly clamped or attached to the piping in a manner that would prevent axial movement of the pipe due to thermal strain. Sometimes piping should be rigidly clamped at valves to prevent continual twisting and pulling at the joint, and recommendations of the pipe manufacturer should be solicited and followed in this regard.

In general, though actual spacing and location of hangers must consider pipe diameter and wall thickness and service conditions, horizontal runs of piping should be supported at intervals of not more than every 8 ft., near the ends of branches, near changes in direction, and near ends of runs in order to provide sufficient rigidity to the system. The actual spacing of supports must limit sagging between supports (as a result of softening or creeping at service temperature under the weight of pipe plus conveyed water) to the extent necessary to avoid the appearance of abnormality in the piping installation. Therefore, support spacings for piping of any diameter used within a building should be based on a maximum allowable sag between supports of 0.5 in. when water within the pipe is at normal maximum service temperature. This requires that spacing of hangers for cold water lines be based on exposure to water at 90° F. while spacing for hot water lines be based on exposure to water at 180° F. In no event, however, should the spacing exceed a maximum of 8 ft. In building areas where ambient temperatures might sometimes be considerably in excess of 90° F. (e.g., in boiler or equipment rooms of multistory structures or unventilated attics of typical single-family residences), hanger spacing for cold water lines should be less than that established based on the water within at a temperature of 90° F. in order to offset any increase in creep or softening of the plastic piping at these higher temperatures. In such cases, a maximum service temperature of 130° F. should be assumed and the required spacing of hangers determined on this basis. Also, it is possible that abusive loads can be applied to exposed piping run horizontally along basement ceilings. Where abuse of significant magnitude can be anticipated, spacing between supports should again be less than that spacing established for prevention of thermal sag alone. Whenever economics permit, horizontal runs of piping in such areas should be concealed or run as close to the ceiling or rafters as possible to reduce the possibility of abusive loads. Recommendations of the pipe manufacturer regarding the spacing required to limit midspan sag of a particular pipe to 0.5 in. at the service temperature should be solicited and followed.

Vertical piping should be supported at a minimum of not less than every story and at its base; for vertical piping of appreciable height (i.e., three stories or more) support with spring hangers is preferred.

3. Insulation

Notwithstanding the advice given in the preceding discussion of thermal expansion and contraction, thermoplastic piping should be insulated when installed in areas where the ambient temperature can be either below 32° F. or above the maximum service temperature of the water being conveyed. The piping should also be insulated or otherwise protected when installed in areas where it is apt to be in close proximity to heat sources such as space heaters, kitchen ovens, and so forth.

When insulation is to be used, the advice of the piping manufacturer should be solicited and followed with respect to the selection of insulation; all systems of insulation may not be compatible with a particular thermoplastic piping material.

4. Underground Burial*

The important considerations in the installation of underground thermoplastic piping are the manner in which external loads are imposed on the piping and the manner in which the piping resists these loads.

External loads (soil burial and surcharge loads) are imposed on underground thermoplastic piping in the same manner that such loads are imposed on underground piping of any other material. In keeping with

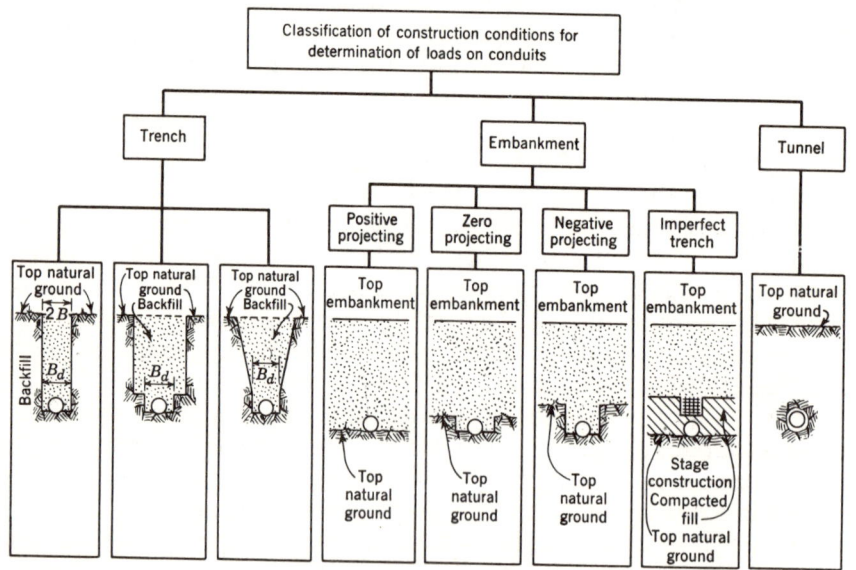

Figure 1. Classification of underground burial conditions. (Courtesy of American Society of Civil Engineers)

---

*See additional remarks in Appendix C by Elmer E. Jones, Jr.

current soil mechanics and engineering practice, the manner in which these loads are imposed is a function of the way in which the piping is installed; for convenience pipe burial conditions are divided into three major classes--trench (ditch) condition, embankment condition, and tunnel condition (Figure 1).

Computation of the amount of the load imposed is readily performed using the formulas, coefficients, settlement ratios, and loading diagrams found in a wide variety of texts and piping handbooks.

The manner in which external loads are resisted by underground piping is largely a function of the tolerance of the piping for diametric deformation. Rigid piping, which cannot tolerate more than 0.1 percent diametric change, resists external loads by development of its ring strength and fails by rupture of the piping wall at the quarter-point when the ring strength is exceeded (see Figure 2).

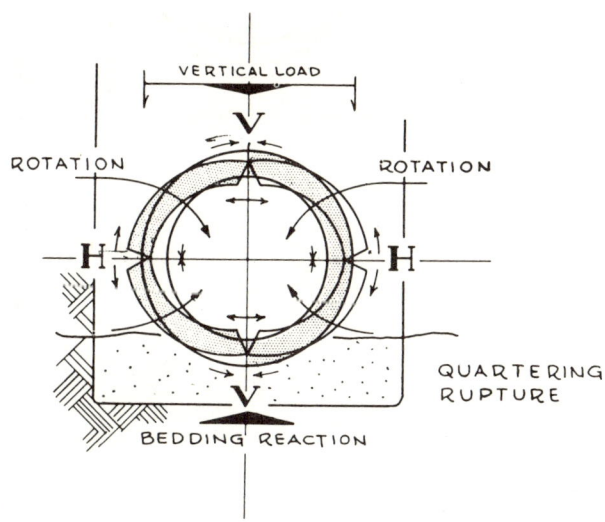

Figure 2. Characteristic failure of externally loaded rigid pipe. (Courtesy of Public Works.)

Flexible piping, which can tolerate more than 3 percent diametric change, resists external loads partly by development of its ring strength and partly by deforming sufficiently to develop the passive soil resistance of side fill material to offset the stresses created by the vertical loadings. Generally speaking, flexible pipe fails when deflection proceeds to such a point that further deflection results and the top of the piping assumes a concave shape that tends to pull the sides of the pipe inward, thus eliminating the passive side resistance of the pipe (see Figure 3).

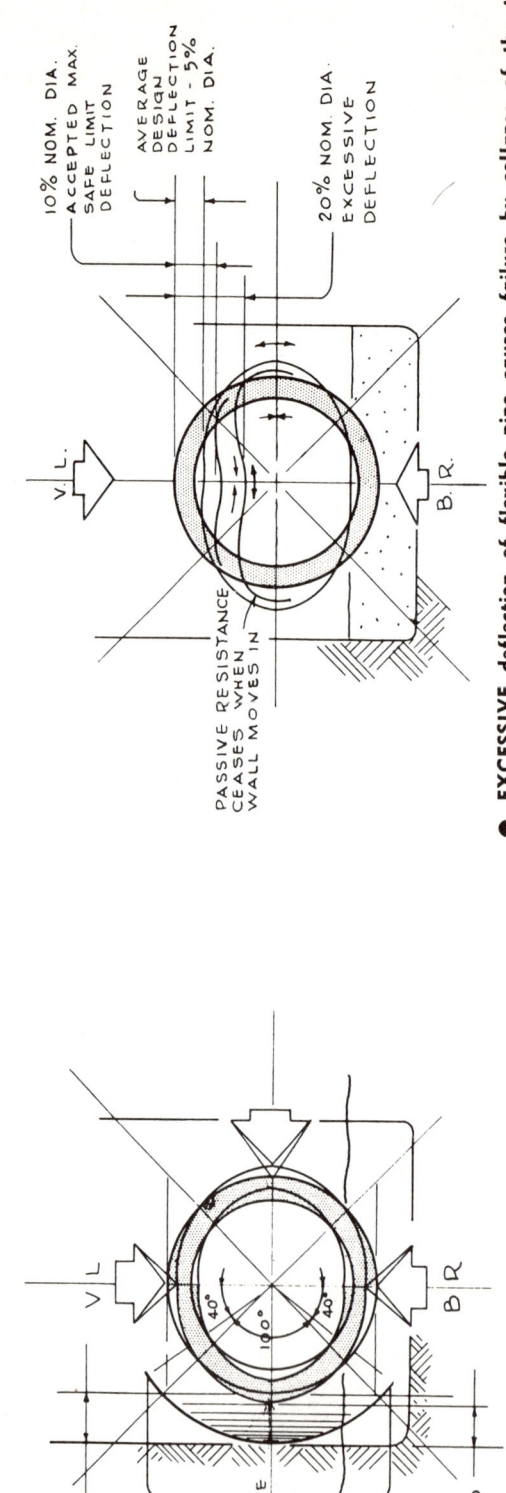

Figure 3. Characteristic failure of externally loaded flexible pipe. (Courtesy of *Public Works*.)

Thermoplastic piping may be classified logically as flexible pipe since it can tolerate more than 3 percent diametric deformation without material damage. At this point, however, the reasonably well defined reactions of flexible piping of nonplastic material no longer consistently apply to thermoplastic piping. Some types of thermoplastic piping fail by crown buckling (as shown in Figure 3), while still other types can be almost completely flattened without failure of the pipe wall. However, in both cases the flexibility of the piping before failure enables it to utilize the passive resistance of the soil to carry a portion of the external (earth, superimposed, and impact) loads imposed upon it, and the loading theories and design practices currently applied to nonplastic flexible piping may be reasonably used by the designer.

However, in applying current loading theories and design practices to thermoplastic piping, selection of values of maximum acceptable deflection should be made not only with a view to preventing pipe failure but also with a view to limiting changes in hydraulic characteristics to a level consistent with the overall system design. Because many of the factors involved in the necessary computation to determine loads are yet to be firmly established, some guidance in this matter is provided below.

With respect to soil loads on piping installed in a trench, when the soil is thoroughly tamped or compacted, the load on thermoplastic piping can be computed in the same way used to determine the load on flexible piping. If, however, thermoplastic piping is laid in a trench and little consideration is given to backfill tamping (as might be the case in some instances for small-diameter water service lines), the load on the thermoplastic piping is certainly somewhat greater than that computed for flexible piping, because the sidefill is then relatively compressible and cannot support much of the load. Therefore, it is recommended that when compaction of the backfill around thermoplastic piping cannot be assured, the value computed for flexible piping should be increased by 50 percent for the thermoplastic piping.

Whenever thermoplastic piping is installed in the trench condition and the trench width (B_d) exceeds the outside diameter of the piping (B_c) by a factor of three or more, the load on the piping should be computed as if the piping were in a positive projecting (embankment) condition.* For those installations where thermoplastic piping is laid in a trench significantly larger than three times the outside diameter with no consideration given to tamping or compacting the backfill, the

---

*Examination of the way in which loads are computed for the trench condition indicates that the load on the pipe is a function of the width of ditch in which the pipe is laid; that is, the wider the ditch, the greater the load on the top of the pipe laid in it. Obviously, there is a limiting width beyond which this principle does not apply, since in a ditch that is very wide relative to the pipe,

load on the piping should be determined as if the piping were in a positive projection embankment condition and the value computed thereby increased by 50 percent.

For thermoplastic piping in a positive projecting (embankment) condition, determination of the load on the piping should be made on the assumption that the settlement ratio has a value of zero. This procedure will result in computation of a reasonably conservative value for the soil load to be experienced, particularly when the backfill soil is tamped.

For thermoplastic piping in a negative projecting (embankment) condition, determination of the load on the piping should be made on the assumption that the settlement ratio has a value of -0.3.

Finally, in the computation of all soil loads, a value of 125 lb. per cu. ft. should be used unless the average maximum weight of the soil which is to be used as backfill or embankment is determined by actual density measurements at an advanced state of design.

From the foregoing, it can be deduced that it is necessary that proper installation techniques be followed to ensure that as much lateral resistance at the sides of the pipe be developed and maintained as is possible. However, it is difficult to ensure that the installation practices set forth below can be followed for small service lines supplying water to one- and two-family residential structures having three stories or less, or for small-diameter pipe in general. Therefore, these installation requirements can be waived for all piping with nominal diameters of less than 5 in., because acceptable pipe has sufficient inherent ring strength or stiffness to offset--without benefit of support from the side soil--the soil burial plus superimposed surface load associated with the particular installation condition without undergoing diametral deflection in excess of 5 percent.

---

the sides will be so far away from the pipe that they cannot possibly affect the load on it. In the case of rigid pipe, as the width of the ditch increases, other factors remaining constant, the load on a rigid pipe increases in accordance with the theory for pipe in a ditch condition until it equals the load determined by the theory for pipe in a projection condition; at greater widths, the load remains constant regardless of width of ditch. Curves have been established which relate values of the ratio of width of ditch to width of pipe (B_d/B_c) at which loads on a rigid pipe are equal when computed with either the ditch or projection condition theory. Unquestionably, the same principle holds true for flexible pipe, but no family of curves relating values for the transition width ratio exist; specific transition width ratios are reported in the National Plumbing Code Handbook for 4-, 6-, and 8-in.-diameter flexible pipe and, under the most adverse conditions, the maximum ratio value is considerably less than 3.

For other than small pipe, ditches (or subditches) should be excavated with straight sides to a smooth bottom and wide enough only to allow for proper joining of pipe lengths, a moderate amount of snaking of the pipe to provide for expansion and contracting with ground and water temperature cycling, and compaction of the backfill soil.

The ditch bottom should be smooth and of uniform grade with a minimum bedding layer of 4 in. prepared with selected fine compactible soil materials free of stones, boulders, waste metal, boards, nails, or other sharp objects that might score, cut, notch, or otherwise damage the pipe by contact or movement. The pipe should be placed in the ditch to bear on the bedding throughout the entire length of its barrel, with offsets shaped in the bedding to fit joint hubs. After the piping is laid, the ditch should be backfilled to a minimum of 12 in. above the top of the piping with the same selected fine compactible soil materials used for the bedding. The backfill should be brought up evenly on both sides of the piping for its entire length, with care taken to compact the backfill under the haunches of the piping to ensure that it is in intimate contact with its sides. All backfill and bedding soil--except for the immediate 12 in. directly above the pipe--should be compacted in layers not exceeding 6 in. in compacted thickness at near-optimum moisture content to a density of at least 90 percent of maximum density at a compactive effort of 12,400 ft.-lb. per cu. ft. The 12 in. directly above the piping should not be compacted to ensure that the soil settles more directly above the pipe than on its sides. Care should be taken to avoid damage to the piping.

When the pipe is installed in a positive projecting (embankment) condition, the natural ground surface should be prepared to fit the pipe, and offsets shaped to fit joint hubs in order to maintain piping true to line and grade. The same type of fine compactible soil materials as used for trench backfilling should be used, and this soil should be compacted on each side of the piping for a width equal to twice the outside diameter for a height of at least 12 in. above the top of the piping. The 12 in. of backfill soil directly above the piping should not be compacted.

Though the inherent flexibility of plastic piping should provide a distribution system with greater resistance to soil movement than that afforded by more rigid piping, when installed in such soils as expansive clays, the practices normally followed to offset soil movement should be adhered to.

Concern with installation cannot end with proper laying and backfilling of the pipeline if a failure-free or leak-free system is to result. That is, adequate attention also must be given to installation and alignment of large valves, curb stops, and service boxes, and to the use of anchors and thrust blocks. The installation and alignment of these items in thermoplastic piping systems should be done in the same way as is done in more conventional piping systems. Recommendations of the pipe manufacturer should be solicited and closely followed in all instances involving installation of such items as large valves and curb boxes and the use of anchorage and thrust blocks, except where such recommendations are contrary to precautions normally employed with regard to installation of piping of other materials of manufacture.

5. <u>Joining</u>

Many methods are available for joining together sections of plastic pipe and for joining plastic pipe to pipe of other materials or to service equipment items. However, for reasons of wall thickness or mechanical and chemical characteristics of the materials themselves, pipe and fittings of some plastics may lend themselves to joining by only one method, while others may permit a choice of several. Of the methods of joining, those discussed in the following paragraphs appear to be the ones most commonly preferred.

The simplest method of joining plastic pipe and fittings employs serrated insert fittings and hose clamps. This method is commonly used to join polyethylene pipe and works best with the low-density type of material, which is soft and flexible. High-density polyethylenes have been successfully joined with insert fittings, though more difficulty must be expected in making tight joints this way because of the greater stiffness of the pipe. To obtain a good seal, the insert fittings must be hard; they are commonly made of acrylonitrile-butadiene-styrene, polyvinyl chloride, nylon, and polypropylene. Stainless steel clamps are generally recommended for buried polyethylene piping systems in order to minimize corrosion.

Like metallic pipe, plastic pipe can also be threaded and this method of joining was very common when plastic piping was first introduced. Threaded connections are useful when the pipeline must be disassembled frequently for cleaning or inspection, but because plastics are notch sensitive, threaded connections are usually recommended only for large pipe schedules (80 and 120). The reduced cross section of threaded schedule-40 pipe and the stress-riser effect of the threads make it more susceptible to failure, particularly on impact; thus it should not be threaded. Threads can be cut in plastic pipe by machine or by hand with standard metal thread-cutting tools, so that it will mate with the standard internal threads of molded fittings and valves. When threading the pipe, however, only thread tape or lubricant recommended by the pipe manufacturer should be used; conventional pipe-thread compounds, putty, linseed-oil-base products, or unknown mixtures should be avoided. Since plastics are relatively flexible, it is considered good practice to insert a long tapered plug in the end during threading to prevent crushing or rupturing of the pipe wall. During the threading operation, it is sometimes necessary to protect the pipe from chuck or vise jaws by pads of leather and rubber or by wrapping the pipe with canvas or emery paper. When assembling a threaded joint, it is first made hand-tight, with additional torque applied by means of a strap wrench. One turn beyond handtight is all that is recommended, because further tightening may split the shoulders of the fitting or cause excessive stress in the joint. Standard pipe wrenches should be avoided, because they will deform and scar the pipe surface. When threaded joints are made between plastic and metal piping, the differences in thermal expansion suggest the use of female plastic components and male metal components to ensure leak tightness. Because of the notch sensitivity, threaded joints should not be used--even for transition joints--if there is an alternative method available.

Solvent-cemented joints using socket or slip-sleeve fittings can be used with pipe made of many different types of thermoplastic materials. Filled solvent cements contain dissolved pipe material; thus it is necessary to use the proper joining compound with a given type of plastic piping.

When solvent-cemented joints are to be used in the piping system, the joints should be made in conformance with recognized current industry practice, such as ASTM D2855-70, Making Solvent-Cemented Joints with Polyvinyl Chloride (PVC) Pipe Fiberglass, and the Plastic Pipe Institute's Recommended Practice for Making Solvent-Cemented Joints with Polyvinyl Chloride Plastic (PVC) Pipe and Fittings.* If such a recognized industry practice does not exist for a particular thermoplastic piping, the joints should be made in conformance with the recommendations of the pipe manufacturer regarding the solvent cement (its specifications, selection, storage), the pipe cleaner or primer, and the procedure to be followed (cutting the pipe, test fitting the joint, joint preparation, application of cement, assembly of joint, set time, and pressure testing).

The advantages of solvent-cemented joints which have led to their widespread use are the ease with which joints can be made and the greater assurance of strong and leak-free joints. Additionally, since the joint does not weaken the pipe as in the case of threading, a thinner wall section can be utilized for a given working pressure. The major drawback of solvent-cemented joints is that they cannot be disassembled. However, if the pipe is compatible with this method of joining, solvent-cementing should be used wherever possible.

With some types of thermoplastic materials (polyethylene or polypropylene, for example), thermal joining techniques such as socket fusion, butt fusion, and hot gas fusion can be used. These methods, which generally require greater skill on the part of the pipefitter than other methods, are fully described in ASTM D2657-67, Recommended Practice for Heat Joining of Thermoplastic Pipe and Fittings.

Flanged joints are also available and are most often used where lines must frequently be disassembled. Flanges are made to standard dimensions to facilitate joining plastic and metallic piping systems and are available commercially with both socket and threaded pipe-to-flange

---

*Plastic Pipe Institute, Recommended Practice for Making Solvent-Cemented Joints with Polyvinyl Chloride Plastic (PVC) Pipe and Fittings - Technical Report PPI-TR10 (New York: Plastic Pipe Institute February 1969).

connections. Gaskets used in flange connections are normally made of neoprene or other rubber-like materials, but only gaskets made of materials compatible with the pipe should be used. The use of a torque wrench for tightening flange bolts is required to ensure that the flange is evenly loaded. Because of creep, however, it may be necessary to check the loading of bolts in flanged joints periodically to ensure tight joints.

From the preceding discussion, it is obvious that consideration must be given to the method of joining to be employed and that the selection of a suitable method of joining plastic piping (when a choice is available for a particular thermoplastic piping) is closely related to whether partial dismantling of the system in the future is contemplated. The pipe manufacturer is best qualified to know the method most appropriate to the particular pipe material, and the manufacturer's recommendations in this regard--as well as those regarding procedures for preparing and making the joints--should be solicited and followed. When given alternatives by the manufacturer, and when design permits, the solvent cementing method of joining piping should be used. When used, only the particular cementing compound recommended by the pipe manufacturer should be used. Solvent cements made in accordance with specifications of the American Society for Testing and Materials should be used when available and they should be stored on the jobsite in containers permanently marked to indicate the solvent cement material, the pipe and fittings with which it may be used, the manner of use, and any safety and storage precautions to be taken in connection with use. Also, when given an alternative, threaded joints should usually be avoided because of the notch sensitivity of plastic; in no case should threaded pipe with wall thickness of less than Schedule 80 be used.

6. <u>Dimensions, Tolerances, and Markings</u>

Joining of one section of pipe to another, of fittings to pipe, of plastic pipe to pipe made of another type of material, of plastic pipe to service equipment, and so forth requires that pipe, fittings, and adapters made of thermoplastic materials meet nationally recognized standards for dimensions and tolerances if hydraulically efficient distribution systems are to be realized.

Because of the availability of numerous polymers (and even more numerous formulations based on each polymer), it is highly desirable that plastic piping be identified with markings indicating the plastic extrusion compound as well as the piping specification designation (e.g., see the marking requirements for polyvinyl chloride in ASTM D1785-68). In addition, to reduce or eliminate many installation problems related to size, origin, and intended use, it is desirable that pipe and fittings for use in potable water distribution systems be further identified with markings indicating nominal inside diameter, pipe schedule or

SDR,* manufacturer's name or trademark, and pressure rating at the intended service temperature (determined in accordance with Test S-1). Additionally, if thermoplastic pipe and fittings are to be installed in more than one piping system within the same building water distribution system, then pipe and fittings either should be so color coded or marked that immediate identification of the proper use of each length of pipe and each fitting is possible, or pipe and fittings for all systems should meet the pipe and fittings criteria for the most severe system.

---

*The standard thermoplastic pipe dimension ratio (SDR) is the ratio of pipe diameter to wall thickness. Thermoplastic piping is commercially available in a range of SDR's for different pipe sizes.

APPENDIX A

EVALUATION PROCEDURES

This appendix presents procedures for evaluating pipe and fittings performance, together with suggested minimum levels of acceptability. Groupings are based on the nature of properties tested--general, structural, chemical, and biological. Unless specifically indicated otherwise, procedures and minimum levels of acceptability apply to all pipe and fittings manufactured from thermoplastic material for use in potable water distribution systems, irrespective of class of service (water mains, water service lines, or building water distribution systems), type of installation (above or below ground), or water conveyed (hot or cold).

GENERAL

TEST G-1: PIPE AND FITTINGS INSPECTION

Purpose

To specify a standard procedure for inspection of pipe and fittings for surface and alignment defects, and to define, by type and number, the defects permissible in pipe and fittings when new and after subjected to other tests.

1. Blister - Undesirable rounded elevation of the surface, whose boundaries may be either more or less sharply defined, somewhat resembling in shape a blister on the human skin. A blister may burst and become flattened.

2. Bubble - Internal void or a trapped globule of air or other gas.

3. Crack - Actual fracture.

4. Crater - Small, shallow, crater-like, surface imperfection.

5. Crazing - Fine cracks at or under the surface.

6. Foreign inclusions - Particles of substance included in a plastic that are foreign to its composition.

7. Pimple - Undesirable, small, sharp, or conical elevation on the surface, whose form resembles a pimple in the common meaning.

8. Pinhole - Very small hole.

9. Pit - Small regular or irregular crater in the surface, usually with its width approximately of the same order of magnitude as its depth.

Procedure

Maintain temperature of test area and pipe and fitting specimens at 73.4° F. (+3.6° F.) and 50 ± 5 percent relative humidity.

Wash specimen with soap and water, and rinse with tap water. Inspect inside and outside surfaces of specimen visually at normal reading distance for blemishes and defects, using as light source either partially diffused daylight or substantially equivalent artificial light, with luminous intensity near inspection surface of not less than 100, nor more than 200, footcandles.

Measure specimen in accordance with ASTM D2122-70, <u>Standard Method of Determining Dimensions of Thermoplastic Pipe</u>, to determine compliance with dimensional requirements in applicable national standards.

Recommended Requirement

The outside and inside walls of the specimen should not contain any of the defects and blemishes defined above. The dimensional properties of the specimen should meet the applicable national standards.

STRUCTURAL

TEST S-1: HYDROSTATIC PRESSURE TEST*

Purpose

To determine whether pipe, fittings, and joints will withstand the potable water system working pressure** and temperature without excessive stresses in the walls of pipe and fittings and without leakage loss in the piping system.

Apparatus and Procedure

The apparatus and procedure are identical to those specified in ASTM D2837-69, <u>Standard Method of Obtaining Hydrostatic Design Basis for Thermoplastic Pipe Materials</u>, except that the procedure is to be modified so that the test specimen consisting of two lengths of pipe and a fitting joined in accordance with the recommendations of the pipe manufacturer is tested.

---

*No standardized test procedure exists. Inasmuch as the collection of test development and evaluation data was beyond the scope of this study, the suggested test merits laboratory verification.

**The potable water system working pressure is the pressure determined by the system designer as being the maximum pressure to be experienced by the system as a result of pumping pressure and pressure associated with anticipated water surges.

Moreover, whereas the temperature of water within the test specimen during the test should be 73° F. for 10,000 hours for pipe fittings, and joints intended to be installed above or below ground and to convey cold water, the temperature of water within the test specimen during the test should be 180° F. for 10,000 hours for pipe, fittings, and joints intended to be installed above or below ground and to convey domestic hot water. Additionally, for pipe, fittings, and joints intended to be installed above or below ground and to convey hot water at a temperature in excess of 180° F. for nondomestic hot water purposes, the test should be performed with the temperature of water within the test specimen at 205° F. ($\pm$5 F.).

Recommended Requirement

For pipe, fittings, and joints intended to be installed above or below ground and to convey hot or cold water, the pressure rating of the test specimens established as a result of this test should exceed the maximum hydrostatic pressure of the system by 50 percent.*

TEST S-2: CRUSHING STRENGTH TEST

Purpose

To determine whether pipe to be installed below ground will withstand the soil burial load plus superimposed surface loads without excess diametral deflection.

Apparatus and Procedure

The apparatus and procedure are identical to those specified in ASTM D2412-68, Standard Method of Test for External Loading Properties of Plastic Pipe by Parallel-Plate Method.

Recommended Requirement

For 6-in. pipe or less, the stiffness factor** determined, at 4 percent deflection, by the test should be equal to or greater than the

---

*Note should be taken that the recommended requirement is stated in terms of the "pressure rating" of the pipe. Calculation of the "pressure rating" of the pipe requires prior determination of the "long-term hydrostatic strength," the "hydrostatic design basis," and the "hydrostatic design stress" in accordance with ASTM D2837-69.

**The stiffness factor is the product of the modulus of elasticity of the pipe material and the moment of inertia of the pipe cross section. The stiffness factor (SF), expressed in pound-inches squared per lineal inch of piping, is normally determined by the stiffness factor (in pound-inches squared per lineal inch of piping length) specified in Table 1.

For pipe larger than 6 in. in diameter, the stiffness factor determined, at 5 percent deflection, by the test should be equal to or greater than the stiffness factor computed in accordance with the following expression:

$$SF = (6W_t - 42.7\ r)\ r^2 \quad \text{(pound-inches squared per lineal inch of piping)}*$$

or

$$SF = (6W_t - 12.2\ r)\ r^2 \quad \text{(pound-inches squared per lineal inch of piping)}**$$

in which W_t is the anticipated soil burial plus superimposed surface loads on the pipe (in pounds per lineal inch of pipe) and r is the nominal radius of the pipe.

---

equation:

$$SF = \frac{D_e W_e K r^3}{\Delta^x} - 0.061\ E'\ r^3,$$

in which D_e is a dimensionless deflection lag factor; W_e is the total intended load (burial load plus superimposed surface load) to be experienced by the pipe, in pounds per lineal inch of piping; K is a dimensionless bedding constant; r is the nominal radius of the pipe in inches; Δ^x is the deflection of the pipe diameter in inches; and E' is the modulus of backfill reaction in pounds per square inch.

For the purpose of this test, the following values of constants should be used:

$D_e = 1.5$

$K = 0.10$

r = one-half the nominal diameter of the pipe

$\Delta^x = 0.025_r$

E' = 700 lb. per sq. in. for piping larger than 6 in. in diameter if installed in accordance with the criteria in Section III; or 200 lb. per sq. in. for all other piping.

*For pipe installed in accordance with installation criteria set forth in Section III.

**For pipe not installed in accordance with installation criteria set forth in Section III.

TABLE 1
RECOMMENDED STIFFNESS FACTORS FOR PIPE 6-IN. OR LESS

Pipe Size (in.)	Stiffness Factor* (lb.-in.^{2} per lineal in.)
1/8	0.23
1/4	0.93
3/8	2.1
1/2	3.8
3/4	8.7
1	16
1-1/4	25
1-1/2	41
2	92
2-1/2	170
3	286
3-1/2	444
4	651
5	1242
6	2106

*The stiffness factor values shown are calculated on the basis of the pipe being in a "projecting conduit" condition and under either 2 or 8 ft. of backfill, depending upon which height produced the greater combined soil burial plus superimposed surface load on the buried pipe.

TEST S-3: CONCENTRATED LOAD TEST*

Purpose

To determine whether pipe, fittings, and joints to be installed above ground will withstand imposed external suspended loads without excessive vertical deflection of the assembly.

Apparatus and Procedure

The apparatus required comprises:

1. A device for supporting the test specimen (an 8-ft. length of pipe, an 8-ft.-long pipe-fitting assembly, and an 8-ft. length of joined pipe) in accordance with recommendations of pipe manufacturer.

2. A device for maintaining water for 10,000 hours in the test specimens at a pressure equal to 80 percent of the pressure rating of the pipe and at a temperature of 73° F. (if the pipe and appurtenances are intended to convey cold water); 180° F. (if the pipe and appurtenances are intended to convey domestic hot water); and 205° F. (if the pipe and appurtenances are intended to convey nondomestic hot water).

3. A device for direct deadweight loading of the test specimen at midspan between two hangers.

4. A device for measuring the vertical deflection at the joint of loading.

The procedure essentially entails installing the test specimen in accordance with recommendations of the pipe manufacturer, sealing the ends of the test specimen, introducing water at the test temperature and pressure into the test specimen, allowing the test specimen to come to temperature equilibrium, measuring the amount of no-load vertical deflection, subjecting the test specimen to a deadweight vertical load (unless unusual loadings are to be anticipated in actual service use, a 25-lb. load is used), measuring the amount of initial vertical deflection, and measuring the amount of subsequent vertical deflections at 200-hour intervals through the 10,000-hour test period.

Recommended Requirement

The total deflection of the test specimen should not exceed 0.3 in. nor should there be any loss of water.

---

*No standardized test procedure exists. Inasmuch as the collection of test development and evaluation date were beyond the scope of this study, the suggested test merits laboratory verification.

CHEMICAL

TEST C-1: WATERBORNE CHEMICAL TEST*

Purpose

To determine whether pipe fittings and joints will withstand chemicals likely to be encountered in conveyed water without damage.

Apparatus and Procedure

The apparatus required comprises:

1. Chemicals listed in Table 2.

2. Eighty-five specimens of 1 1/2-in. piping, each 12 in. long and each comprising two lengths of pipe and a fitting, joined in accordance with recommendations of the pipe manufacturer.

3. A device for maintaining test specimens full of the chemicals listed in Table 2 for a period of 240 hours at a temperature of 73° F. (if the piping is intended to convey cold water); at 180° F. (if the piping is intended to convey domestic hot water); and at 205° F. (if the piping is intended to convey nondomestic hot water), and at the pressure rating of the piping as determined by Test S-1.

4. Apparatus specified in ASTM D1599-69, Standard Method of Test for Short-Time Rupture Strength of Plastic Pipe, Tubing, and Fittings.

Prepare standard solutions of the chemicals list in Table 2 as follows:**

1. Alkyl Benzene Sulfonate (ABS). Obtain ABS reference material from The Soap & Detergent Association (40 E. 41st Street, New York, N.Y. 10017). Weigh an amount of the reference material equal to 1.000 g ABS on a 100 percent active basis. Dissolve in distilled water

---

*No standardized test procedure exists. Inasmuch as the collection of test development and evaluation data were beyond the scope of this study, the suggested test merits laboratory verification.

**The preparations, except for carbon chloroform extract, are identical to those set forth in American Public Health Association, American Water Works Association, and Water Pollution Control Federation, Standard Methods for the Examination of Water and Wastewater, 12th Edition (New York: American Public Health Association, Inc., 1965) as the standard solutions to be used in performing analyses of water. The method of preparing the carbon chloroform extract solution is based upon using a reference material obtained by the Environmental Protection Administration by the "Carbon Chloroform Extract Method" tentatively recommended in Standard Methods.

and dilute to 1,000 ml. Dilute 10.00 ml of this stock solution to 1,000 ml with distilled water (1.00 ml = 10.0 micrograms ABS).

2. Arsenic (As). Dissolve 1.320 g As_2O_3 in 10 ml distilled water containing 4 g NaOH, and dilute to 1,000 ml with distilled water. Dilute 5.00 ml of this stock solution to 500 ml with distilled water (1.00 ml = 10.0 micrograms As).

3. Barium (Ba). Dissolve 35.6 g $BaCl_2 \cdot 2H_2O$ in distilled water, and dilute to 1,000 ml with distilled water. Dilute 100 ml of this stock solution to 1,000 ml with distilled water (1.00 ml = 2 mg Ba).

4. Cadmium (Cd). Dissolve 0.1000 g pure Cd metal in a solution composed of 20 ml distilled water plus 5 ml concentrated HCl. Use heat to assist the dissolution of the metal. Transfer the solution quantitatively to a one-liter volumetric flask, and dilute to the mark with distilled water. Pipet 10.00 ml of this stock solution into a one-liter volumetric flask, add 10 ml concentrated HCl, and dilute to the mark with distilled water (1.00 ml = 1.00 micrograms Cd).

5. Carbon Chloroform Extract (CCE). Obtain CCE reference material from the Environmental Protection Administration (Division of Water Hygiene, Cincinnati Laboratories, 5555 Ridge Avenue, Cincinnati, Ohio 45213). Weigh an amount of the reference material equal to 1.000 g CCE on a 100 percent active basis. Dissolve in distilled water and dilute to 1,000 ml. Dilute 10.00 ml of this stock solution to 1,000 ml with distilled water (1.00 ml = 10.0 micrograms CCE).

6. Chloride (0.0141N NaCl). Dissolve 0.8241 g NaCl (dried at 140° C.) in chloride-free water, and dilute to 1,000 ml (1.00 ml = 0.500 mg Cl^-).

7. Chlorine (OCl^-). Dilute Zonite (Zonite Products Corporation) or household bleach with chlorine-demand-free water to obtain approximately 250 mg available chlorine per liter (Zonite contains approximately 1 percent available chlorine). Titrate this stock solution with 0.01N or 0.025N $Na_2S_2O_3 \cdot 5H_2O$ to obtain 250 mg per liter of available chlorine. Note: all glassware and water used in the preparation of this standard solution must be chlorine-demand free.

8. Chromium, Hexavalent (Cr^{+6}). Dissolve 0.1414 g $K_2Cr_2O_7$ in distilled water, and dilute to 1,000 ml. Dilute 20.00 ml of this stock solution to 1,000 ml with distilled water (1.00 ml = 1 microgram Cr^{+6}).

9. Copper (Cu). Dissolve 0.1000 g Cu metal foil in a solution composed of 3 ml copper-free water and 3 ml concentrated HNO_3. After the metal has dissolved, add 1 ml concentrated H_2SO_4 and heat to volatilize the acids. Stop heating just short of complete dryness. Cool and dissolve in copper-free water. Transfer quantitatively to a one-liter volumetric flask, and make up to the mark with copper-free water (1.00 ml = 0.100 mg Cu).

10. Cyanide (CN). Dissolve 2.51 g KCN in one liter of water. Standardize this solution against 0.0192N $AgNO_3$ to obtain approximately 1 mg CN per ml stock solution. Dilute 10 ml of this stock solution to 1,000 ml with distilled water; mix, and make a second dilution of 10 ml to 100 ml with distilled water (1.00 ml = 1.0 microgram CN).

11. Fluoride (F). Dissolve 0.2210 g NaF in distilled water, and dilute to 1,000 ml. Dilute 100.0 ml of this stock solution to 1,000 ml with distilled water (1.00 ml = 10.0 micrograms F).

12. Iron (Fe). In a one-liter volumetric flask, dissolve 0.2000 g electrolytic iron wire in 20 ml 6N H_2SO_4, and dilute to the mark with iron-free distilled water. Pipet 50.00 ml of this stock solution into a one-liter volumetric flask and dilute to the mark with iron-free distilled water (1.00 ml = 10.0 micrograms Fe).

13. Lead (Pb). Dissolve 1.599 g $Pb(NO_3)_2$ in lead-free water to which has been added 1 ml concentrated HNO_3, and dilute to 1,000 ml. Dilute 10.00 ml of this stock solution to 200 ml with lead-free water; mix, and make a second dilution of 10.00 ml to 250 ml with lead-free water (1.00 ml = 2.00 micrograms Pb).

14. Manganese (Mn). Dissolve 3.2 g $KMnO_4$ in distilled water and dilute to one liter. Age solution in sunlight for several weeks or heat for several hours near the boiling point. Filter solution and standardize against $Na_2C_2O_4$. Calculate the volume of this solution necessary to prepare one liter of solution containing 0.050 mg Mn per 1.00 ml, and prepare this volume. To this volume add 2 to 3 ml concentrated H_2SO_4 and then sodium bisulfite solution (10 g $NaHSO_3$ plus distilled water) dropwise until the permanganate color disappears. Boil to remove excess SO_2, cool, and dilute to 1,000 ml with distilled water. Dilute further if necessary.

15. Nitrate (NO_3^-). Dissolve 0.7218 g KNO_3, and dilute to 1,000 ml with distilled water. Evaporate 50.0 ml of this stock solution to dryness; dissolve the residue by rubbing with 2.0 ml $C_6H_3OH(SO_3H)_2$ reagent, and dilute to 500 ml with distilled water (1.00 ml = 44.3 micrograms NO_3^-).

16. Phenol. Dissolve 1.00 g reagent-grade phenol in freshly boiled and cooled distilled water and dilute to 1,000 ml. Dilute 10.0 ml of this stock solution to 1,000 ml with freshly boiled and cooled distilled water. Dilute 50.0 ml of this intermediate solution to 500 ml with freshly boiled and cooled distilled water (1.0 ml = 1.0 micrograms phenol).

17. Selenium (Se). Place an accurately weighed pellet of ACS-grade metallic Se into a small beaker. Add 5 ml concentrated HNO_3. Warm until the reaction is complete and cautiously evaporate just to dryness. Dilute to 1,000 ml with distilled water and calculate concentration of Se per 1,000 ml in stock solution. Dilute an appropriate volume of this stock solution with distilled water so that 1.00 ml = 1.00 micrograms Se.

18. Silver (Ag). Dissolve 0.1260 $AgNO_3$ in distilled water and dilute to 1,000 ml. Dilute 5.00 ml of this stock solution to 1,000 ml with distilled water. Store overnight to allow the plating-out process to reach equilibrium. Discard this weak solution and make again in the same flask just before using (1.00 ml = 0.40 micrograms Ag).

19. Sulfate (SO_4^{-2}). Dissolve 0.1479 g Na_2SO_4 in distilled water. Dilute to 1,000 ml (1.00 ml = 0.10 micrograms SO_4^{-2}).

20. Zinc (Zn). Dissolve 0.1000 g 30-mesh Zn metal in a slight excess of 1:1 HC; about 1 ml is required. Dilute to 1,000 ml with zinc-free water. Dilute 10.00 ml of this stock solution to 1,000 ml with zinc-free water (1.00 ml = 1.00 micrograms Zn).

Clean 80 specimens with mild detergent, rinse thoroughly with water, and dry. Fill four specimens with one of the test chemicals and maintain the specimens full of the chemical for a period of 240 hours at the stipulated temperature and pressure. Fill another four specimens with another one of the test chemicals and maintain the specimens as indicated for the first four specimens. Continue the procedure, using four specimens for each of the chemicals listed in Table 2.

At the end of the 240-hour period, test each of the 80 specimens in accordance with the procedures of ASTM D1599-69 and calculate the hoop stress in the pipe specimens at failure.

Test the five specimens not subjected to the chemicals in accordance with the procedures of ASTM D1599-69 and calculate the hoop stress in the pipe specimens at failure.

Recommended Requirement

The hoop stress at failure of the test specimens subjected to the chemicals should be equal to or greater than the hoop stress at failure of the test specimens not subjected to the chemicals.

TABLE 2
CHEMICALS FOR USE IN WATERBORNE CHEMICAL TEST

No.	Principal Chemical Ingredient	Concentration of Principal Ingredient (milligrams per liter)
1	Alkyl benzene sulfonate	0.5
2	Arsenic	0.05
3	Barium	1.
4	Cadmium	0.01
5	Carbon chloroform extract	0.2
6	Chloride	250.
7	Chlorine	0.5
8	Chromium	0.05
9	Copper	1.
10	Cyanide	0.2
11	Fluoride	2.4
12	Iron	0.3
13	Lead	0.05
14	Manganese	0.05
15	Nitrate	45.
16	Phenols	0.001
17	Selenium	0.01
18	Silver	0.05
19	Sulfate	250.00
20	Zinc	5.

TEST C-2: WATER CONTAMINATION TEST

Purpose

To determine whether pipe, fittings, and joints will decrease the potability of water being conveyed, below minimum levels recommended by the U. S. Public Health Service.*

Apparatus and Procedure

The apparatus and procedure are identical to those reported in National Sanitation Foundation, A Study of Plastic Pipe for Potable Water Supplies, 1955

Recommended Requirement

After the test, the water should meet minimum requirements of the U. S. Public Health Service, and should not contain any arsenic, barium, cadmium, chromium, chronium (hexavalent), cyanide, fluoride, lead, selentium, or silver above the concentration in the water prior to test.

TEST C-3: WEATHERING RESISTANCE TEST

Purpose

To determine whether pipe and fittings to be installed above ground will withstand ambient weather conditions without damage.

Apparatus and Procedure

The apparatus and procedure are identical to those contained in (a) Procedure C of ASTM D1501-65T, Tentative Recommended Practice for Exposure of Plastic to Flourescent Sunlamp and (b) ASTM D790-66, Standard Method of Test for Flexoral Properties of Plastics.

Make specimens--cut from flat sheets, plates, or molded shapes of the pipe or fitting compound, as molded to the required dimensions and including all stabilizers, lubricants, dyes, pigments, and fillers normally included in the finished product--0.125 in. thick, with other dimensions conforming to those specified in ASTM D790.

Recommended Requirement

Specimens tested in accordance with ASTM D1501-65T should have no defects when examined in accordance with Test G-1: Pipe and Fitting Inspection and, additionally, should not undergo changes in flexural strength and modulus of elasticity in flexure in excess of 10 percent when compared with strength and modulus values of similar exposed specimens.

---

*U.S. Department of Health, Education, and Welfare, Public Health Service, Public Health Service Drinking Water Standards (Washington: U.S. Government Printing Office, latest edition).

BIOLOGICAL

TEST B-1: FUNGUS GROWTH RESISTANCE TEST

Purpose

To determine whether pipe and fittings will sustain microorganism growth.

Apparatus and Procedure

Test organism, culture medium, and inoculum are required as follows:

- a. Test organism used in this test is Aspergullus niger, ATCC No. 6275.* Carefully maintain organisms on potato-dextrose agar medium and promptly renew in event of evidence of contamination. Do not keep stock culture for more than 4 months; keep only in refrigeratio at temperature of approximately 37° to 50° F. Use subcultures incubated at 82° to 86° F. for 10 to 14 days for preparing inoculum.

- b. Prepare culture medium by mixing the following ingredients with distilled water to make 1,000 ml of solution:

$NaNO_3$	3.00 gm
K_2HPO_4	1.00 gm
$MgSO_4 \cdot 7H_2O$	0.50 gm
KCI	0.25 gm
Agar	15.00 gm

    Adjust pH value of culture medium to 5.5 to 6.5 by addition of HCL or NaOH. After mixing, sterilize culture medium by autoclaving for 15 minutes at 15 psi (121° C.). Under sterile conditions, pour the medium into 10-cm petri dishes, about 30 ml per dish, and allow to harden.

- c. Prepare inoculum by adding 10 ml of sterile distilled water containing about 0.005 percent of nontoxic wetting agent to subculture (10 to 14 days old) of test organism in ripe fruiting condition. Force spores into suspension with sterile camel-hair brush (or other suitable means) and dilute to 100 ml with sterile distilled water.

---

*This organism may be obtained from American Type Culture Collection, 12301 Parklawn Drive, Rockville, Maryland 20825 or, if for Department of Defence use, from Pioneering Research Division, U. S. Army Natick Labs, Natick, Massachusetts 01762.

Condition three test specimens—each 1.5-in. squares cut from flat sheets, plates, or molded shapes of the pipe or fitting compound, or molded to the required dimensions and including all stabilizers, lubricants, dyes, pigments, and fillers normally included in the finished products—for a minimum of 48 hours in an area at 73° F. ± 2° and 50 ± 5 percent relative humidity.

Under aseptic conditions, immerse test specimens in 70 percent alcohol for a few seconds, rinse thoroughly in distilled water, then lay flat on furface of hardened medium, one specimen to each dish. With sterile pipet, distribute 1.0 to 1.5 ml of inoculum over surface of the specimen and surrounding medium and incubate for 14 days at temperature of 82° to 86° F. and 85 to 90 percent relative humidity.

Inoculate solidified potato-dextrose agar medium in three separate petri dishes with test organisms to determine viability of inoculum. At end of incubation period, surface of hardened medium in each control dish should be covered with fungus growth.

Upon completion of insulation period, examine test specimens visually.

Recommended Requirement

Tested specimens should show no evidence of fungus growth. If one of three tested specimens shows evidence of fungus growth, test should be performed again on three new specimens. If one of three tested specimens again shows evidence of fungus growth, pipe and fittings should be deemed unacceptable.

APPENDIX B

STANDARDS FOR PLASTIC PIPING

This appendix presents a listing of the standards for thermoplastic piping of the American National Standards Institute and the American Society for Testing and Materials.* Included within the listing are component specifications, methods of test, and recommended practices for pipe, fittings, and related products.

AMERICAN NATIONAL STANDARDS INSTITUTE, INC.**

ANS B16.27-1962	Plastic Insert Fittings for Flexible Polyethylene Pipe
ANS B72.1-1967	Specification for Polyethylene (PE) Plastic Pipe (SDR-PR) (ASTM D 2239-67)
ANS B72.2-1967	Specification for Polyvinyl Chloride (PVC) Plastic Pipe (SDR-PR) (ASTM D 2241-67)
ANS B72.3-1967	Specification for Acrylonitrile-Butadiene-Styrene (ABS) Plastic Pipe (SDR-PR and Class T) (ASTM D 2282-66)
ANS B31.8-1968	USA Standard Code for Pressure Piping, Gas Transmission and Distribution Piping Systems

AMERICAN SOCIETY FOR TESTING AND MATERIALS***

D 1503-68	Cellulose Acetate Butyrate (CAB) Plastic Pipe, Schedule 40
D 1527-69	Acrylonitrile-Butadiene-Styrene (ABS) Plastic Pipe, Schedule 40 and 80
D 1598-67	Time-to-Failure of Plastic Pipe Under Long-Term Hydrostatic Pressure

---

*The listing is based upon Plastic Pipe Institute, Standards for Plastic Piping - Technical Report PPI-TR5-SEP1969 (New York: Plastic Pipe Institute, 1969) and the 1 March 1970 Supplement thereto.

**1430 Broadway, New York, N.Y. 10018.

***1916 Race Street, Philadelphia, Pa. 19103.

D 1599-69	Short-Time Rupture Strength of Plastic Pipe, Tubing, and Fittings
D 1785-68	Poly(Vinyl Chloride) (PVC) Plastic Pipe, Schedules 40, 80 and 120
D 2104-68	Polyethylene (PE) Plastic Pipe, Schedule 40
D 2122-67	Determining Dimensions of Thermoplastic Pipe
D 2152-67	Quality of Extruded Poly(Vinyl Chloride) Pipe by Acetone Immersion
D 2153-67	Calculating Stress in Plastic Pipe Under Internal Pressure
D 2235-67	Solvent Cement for Acrylonitrile-Butadiene-Styrene (ABS) Plastic Pipe and Fittings
D 2239-67	Polyethylene (PE) Plastic Pipe (SDR-PR)
D 2241-67	Poly(Vinyl Chloride) (PVC) Plastic Pipe (SDR-PR and Class T)
D 2282-69a	Acrylonitrile-Butadiene-Styrene (ABS) Plastic Pipe (SDR-PR and Class T)
D 2290-64T	Apparent Tensile Strength of Parallel Reinforced Plastics by Split-Disk Method
D 2321-67	Underground Installation of Flexible Thermoplastic Sewer Pipe
D 2412-68	External-Loading Properties of Plastic Pipe by Parallel-Plate Loading
D 2444-67	Impact Resistance of Thermoplastic Pipe and Fittings by Means of a Tup (falling weight)
D 2446-68	Cellulose Acetate Butyrate (CAB) Plastic Pipe (SDR-PR) and CAB Plastic Tubing
D 2447-68	Polyethylene (PE) Plastic Pipe, Schedules 40 and 80 Based on Outside Diameter
D 2464-67	Threaded Poly(Vinyl Chloride) (PVC) Plastic Pipe Fittings, Schedule 80
D 2465-68	Threaded Acrylonitrile-Butadiene-Styrene (ABS) Plastic Pipe Fittings, Schedule 80
D 2466-67	Socket-Type Poly(Vinyl Chloride) (PVC) Plastic Pipe Fittings, Schedule 40
D 2467-67	Socket-Type Poly(Vinyl Chloride) (PVC) Plastic Pipe Fittings, Schedule 80
D 2468-68	Socket-Type Acrylonitrile-Butadiene-Styrene (ABS) Plastic Pipe Fittings, Schedule 40
D 2469-68	Socket-Type Acrylonitrile-Butadiene-Styrene (ABS) Plastic Pipe Fittings, Schedule 80
D 2513-68	Thermoplastic Gas Pressure Pipe, Tubing, and Fittings
D 2560-67	Solvent Cements for Cellulose Acetate Butyrate (CAB) Plastic Pipe and Fittings
D 2564-67	Solvent Cements for Poly(Vinyl Chloride) (PVC) Plastic Pipe and Fittings
D 2609-68	Plastic Insert Fittings for Polyethylene (PE) Plastic Pipe
D 2610-68	Butt Fusion Polyethylene (PE) Plastic Pipe Fittings, Schedule 40
D 2611-68	Butt Fusion Polyethylene (PE) Plastic Pipe Fittings, Schedule 80

D 2657-67	Heat Joining of Thermoplastic Pipe and Fittings
D 2661-68	Acrylonitrile-Butadiene-Styrene (ABS) Plastic Drain, Waste, and Vent Pipe and Fittings
D 2662-68	Polybutylene (PB) Plastic Pipe (SDR-PR)
D 2665-68	Poly(Vinyl Chloride) (PVC) Plastic Drain Waste, and Vent Pipe and Fittings
D 2666-67T	Polybutylene (PB) Plastic Tubing
D 2672-68a	Bell End PVC Plastic Pipe
D 2683-68T	Socket-Type Polyethylene (PE) Fittings for SDR 11.0 PE Pipe
D 2729-68	Polyvinyl Chloride (PVC) Sewer Pipe and Fittings
D 2737-68T	Polyethylene (PE) Plastic Tubing
D 2740-68	Polyvinyl Chloride (PVC) Plastic Tubing
D 2749-68	Standard Definitions of Terms Relating to Plastic Pipe Fittings
D 2750-69	Acrylonitrile-Butadiene-Styrene (ABS) Plastic Utilities Conduit and Fittings
D 2751-69	Acrylonitrile-Butadiene-Styrene (ABS) Sewer Pipe and Fittings
D 2774-69	Recommended Practice for Underground Installation of Thermoplastic Pressure Piping
D 2837-69	Obtaining Hydrostatic Design Basis for Thermoplastic Pipe Materials
D 2846-69T	Chlorinated Polyvinyl Chloride (CPVC) Plastic Hot Water Distribution Systems
D 2855-70	Making Solvent-Cemented Joints with Polyvinyl Chloride (PVC) Pipe and Fittings

APPENDIX C

ADDITIONAL REMARKS ON UNDERGROUND INSTALLATION

by Elmer E. Jones, Jr.

Three properties of thermoplastic pipe that require special consideration for underground installation are its lightweight, its low beam strength, and its increasing plasticity with increasing temperature. Because of these three properties, those unfamiliar with plastic pipe may unintentionally stress a pipe to the point where failure occurs.

In trying to make a neat, straight installation of plastic pipe under the hot sun, it is possible to stretch the pipe so that 100 feet of pipe is placed in 102 feet of trench. This error can be compounded by improper tamping that lifts the pipe from the trench bottom in an irregular pattern. Six-inch 160-psi pipe weighs only 3.4 pounds per foot. Extremely low improperly applied tamping forces can cause uplift.

The U. S. Department of Agriculture has found the following general procedure satisfactory.

The trench bottom should be uniform and clean and free of objects or rock projections that would injure the pipe surface. Pipe should be snaked (side to side) in the trench such that 100 feet of pipe occupies about 98 feet of trench. If possible, water should be run through the pipe to cool it to or below design soil and water temperature. The pipe should be pressure tested and permitted to remain under pressure during backfill. Carefully selected loose granular backfill capable of easy compaction should be placed in the trench to a minimum depth of 6 in. over the pipe or twice the pipe diameter. Vibrating tampers that will cause the material to flow around the pipe are preferred. In their absence, broadfaced tampers that will not subject the pipe to concentrated local impact should be used. As with any new engineering material gaining wider acceptance, improved procedures will come with experience. Every effort should be made to profit from the advice of the fabricators' technical services.

Pipe installations are made in many different subsurface conditions. It is extremely important to evaluate these and adjust installation procedures to suit local conditions.

External loads are of concern for buried pressure piping, primarily only during the period of construction, placement, cover, and compaction of backfill prior to pressurization of the line. Once the line is pressurized, internal forces will normally greatly exceed external forces. Thirty psi, a low working pressure, is 4,320 psf. It is important to realize that plastic pipe is the most flexible of flexible pipe. Most other flexible pipe has a basic ring for strength, with an inner and outer protective coating. The deflection limit is often determined by the flexibility of the protective coatings.

The following is based on Section 8.27, AWWA Steel Pipe Manual M11, 1964. The practice of determining flexible-pipe wall thickness on the basis of yield strength to resist internal pressure is sound and conservative. Theoretical bending stresses and tensile stress should not be totaled for flexible pipe as must be done for rigid pipe.

Buried pipe when empty is under compressive forces due to soil pressure. The external vertical forces tend to flatten the pipe and create tensile and compressive stresses in the pipe wall owing to a decrease in vertical diameter. When internal pressure is introduced into the pipe, the first action is reversed and the internal pressure tends to force the pipe to a perfect circle. The internal pressure reduces the stresses caused by the soil. In effect, soil pressure prestresses the flexible pipe to reduce hoop stress in the pipe wall. Whatever compressive stress due to soil action that exists in the pipe must be overcome by the tensile force due to internal pressure before the pipe wall can go into tension.

For further information see ASTM D-2774-69, <u>Recommended Practice for Underground Installation of Thermoplastic Pressure Piping</u>.

TH
7
F4
#61

MAR 31 1972

RAYMOND H. FOGLER LIBRARY
DATE DUE

JAN 8 1973

DEC 2 8 1977

DEC 1 8 1981